Outsmart
Waste

Outsmart Waste

The Modern Idea of Garbage
and How to Think Our Way Out of It

Tom Szaky

BK

Berrett–Koehler Publishers, Inc.
San Francisco
a BK Currents book

Berrett-Koehler Publishers, Inc.
235 Montgomery Street, Suite 650, San Francisco, CA 94104-2916
Tel: (415) 288-0260   Fax: (415) 362-2512   www.bkconnection.com

Ordering Information

Quantity sales. Special discounts are available on quantity purchases by corporations, associations, and others. For details, contact the "Special Sales Department" at the Berrett-Koehler address above.

Individual sales. Berrett-Koehler publications are available through most bookstores. They can also be ordered directly from Berrett-Koehler: Tel: (800) 929-2929; Fax: (802) 864-7626; www.bkconnection.com.
Orders for college textbook/course adoption use. Please contact Berrett-Koehler:
Tel: (800) 929-2929; Fax: (802) 864-7626.
Orders by U.S. trade bookstores and wholesalers. Please contact Ingram Publisher Services, Tel: (800) 509-4887; Fax: (800) 838-1149; E-mail: customer.service@ingrampublisherservices.com; or visit www.ingrampublisherservices.com/Ordering for details about electronic ordering.

Berrett-Koehler and the BK logo are registered trademarks of Berrett-Koehler Publishers, Inc.

Printed in the United States of America

Berrett-Koehler books are printed on long-lasting acid-free paper. When it is available, we choose paper that has been manufactured by environmentally responsible processes. These may include using trees grown in sustainable forests, incorporating recycled paper, minimizing chlorine in bleaching, or recycling the energy produced at the paper mill.

Library of Congress Cataloging-in-Publication Data
Szaky, Tom.
   Outsmart waste : the modern idea of garbage and how to think our way out of it / by Tom Szaky.
      pages cm
   Includes bibliographical references and index.
   ISBN 978-1-62656-024-6 (pbk.)
1. Refuse and refuse disposal. 2. Waste minimization. 3. Recycling (Waste, etc.)
I. Title.
   TD791.S94 2014
   363.72'8—dc23
                                                            2013034263

18   17   16   15   14   10   9   8   7   6   5   4   3   2   1

Cover design by M.80/Wes Youssi.
Composition by Gary Palmatier, Ideas to Images. Elizabeth von Radics, copyeditor; Mike Mollett, proofreader; Alexandra Nickerson, indexer.

To Edith and Martin Stein

Contents

Foreword

Waste is natural to every living system. We all consume and at some point eliminate. Eventually, everything and everyone has an end of life.

We may frown on someone who litters or tosses a cigarette butt on the street, but is putting a candy wrapper in the garbage bin—only for it to be trucked to a landfill—much better for the planet? With the more than 4 pounds of garbage the average American discards every day, our individual contributions to this collective trove of waste are hard to countenance, and, largely, we don't.

Globally, humanity has evolved into the modern disposable society, readily buying and discarding non-recyclable products and packaging that were designed to enhance consumer convenience and regular repeat purchases. For the most part, consumers—and I am very much a part of this—buy disposable items and discard them, and we are largely inattentive to where our waste goes. With an addictive satisfaction (but largely without conscience), we each contribute to vast concentrations of waste that nature can't digest and that add toxins into the atmosphere. Our blind eye to how much waste we produce and contribute is a spiritual breach. The flush syndrome is very much an aspect of our shadow, and it festers into a collective disrespect for our home—our planet.

Tom Szaky is a waste pioneer and an eco-capitalist. At age 31 he's running a company that operates in 24 countries, collecting and recycling waste that is otherwise landfilled or incinerated. In this book Tom illuminates pathways to finding "gold in garbage heaps" and, more importantly, explains how human-created waste can be reused, recycled, and reintegrated into our commercial systems. Through his company, TerraCycle, and this book, Tom is tackling a seemingly unsolvable global problem to which each individual contributes. Thanks to this book, I can no longer acquire and discard unconsciously, and as I've long said, change begins with awareness.

Tom's prescription isn't abstinence: He too likes to buy and own, and he is very much aware of the short life cycle of most goods. Rather, Tom suggests that we might consume our way out of the problem—practically aligning the economic forces that drive consumerism to a positive role in the solution. Tom brings the global waste picture into new focus, and in so doing he may help us solve the individual and societal compromises we each make when we somewhat blindly and seemingly helplessly discard and pollute our planet, our home, and our corporal body.

—Deepak Chopra

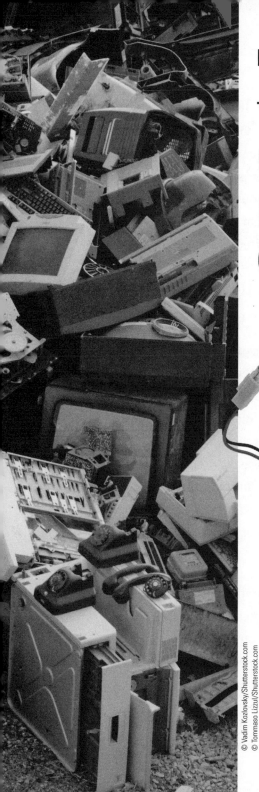

Introduction

The Unique Nature of Garbage

Garbage" is a uniquely human concept that does not exist in nature. In nature the output, or waste, of one organism is the useful input for other organisms. Feces from a fox can become food for a berry bush, whose fruit can later become the food for a bird that may end up as supper for the fox whose droppings started it all. This natural harmony is rooted in the principle that the outputs of organisms tend to bring significant, if not fundamental, benefits to other organisms.

With the creation of synthetic materials, humans have broken this natural harmony. While plastics and other man-made materials have allowed us to innovate and create products cheaply, when they hit the end of their useful life they become useless outputs that nature doesn't know what to do with. Not only are many of these new products relatively cheap to buy but many of us typically don't even have to have the actual resources to buy them; gaining debt (through credit cards and other loans) is perhaps the easiest it has ever been.

Of course, there are ways to better realign ourselves with the harmony of nature. Buying products differently—buying consciously, buying durable, buying used, or simply not buying at all—is a straightforward way that individual consumption can have a smaller impact on nature.

It is quite difficult, however, to lead a life in which we do not buy anything or buy only our bare essentials (food and a few scraps of fabric to cover our bodies). I have started down the path of rethinking what I buy and have found it to be an uphill battle. Like most people, I enjoy acquiring things; the feeling when I open a box with something new to possess inside it is still thrilling, and that fleeting thrill is encouraged by a global culture of rampant consumerism. Just think of how many stores and advertisements we pass by on a daily basis that encourage more and more consumption—all seeming to scream, "You'll gain happiness by buying me!"

We see fish with bellies full of plastic and birds making nests from cigarette butts, and the problem only compounds with our tendency to overconsume. Easy and cheap access to many goods, a dramatic increase in global population, and a throwaway consumer culture have resulted in a global garbage crisis.

What Currently Happens to Our Waste?

Our waste is a monumental problem. Over the past 100 years, the amount of waste that humanity produces has increased by almost 10,000 percent. Developed countries produce more than 4 pounds of waste per person per day.[1] Of that staggering volume, it is estimated that 25 percent ends up in our oceans, forming five gigantic, Texas-sized ocean gyres of garbage.[2] Because of the complexity of much of our garbage, only a small percentage gets recycled.

The majority of the waste that isn't recycled and doesn't wind up in the ocean is effectively mummified and compressed in landfills, leaching out methane and other toxic outputs over time. If it is not buried, it is typically burned in incinerators. While a very small percentage of incinerators do produce some energy as an output, in the process they also destroy all possible value except the caloric (or energy) value inherent in the materials. You can burn something only once.

While the global garbage crisis touches every individual in the world and grows every year, there is cause for optimism. Garbage is a rare example of an environmental problem over which, as individuals, we have tremendous control. The key question is: why do we spend huge amounts of resources—energy, money, and time—to extract oil from the ground and refine it into high-grade plastics, only to burn or bury it after one short use?

Unfortunately, and unlike nature, we often view our waste as something without any inherent value. Fortunately, it doesn't have to be this way.

A Circular Solution to Waste

To properly manage our waste, we need to bring a perspective of value to it, as nature does. Instead of seeing waste purely as a negative—a useless by-product that we spend money to burn or bury—perhaps we can start seeing it as a positive: an inherently valuable combination of materials that can be processed and shaped into objects with specific purposes. The key is to see our outputs not as problems but as assets; it is to see "waste" not as the end of a linear process but as a stage in a circular life cycle.

Reuse—a synonym of *buying used*—is perhaps the solution that most clearly sees the value inherent in our waste. It effectively says that the "waste" object is waste only in the eyes of the initial user; the object retains all of its initial utility in the eyes of the next user and because of that perspective doesn't actually end up as "garbage." If I'm tired of my jeans and put them in the local clothing drop and someone else buys them a few months later, that pair of pants was never rendered waste: they didn't end up in a landfill, and a new pair of pants did not have to be made to meet the needs or desires of the second user.

Not everything is as simple to reuse as a pair of jeans, and most human waste cannot be reused at all. From an empty potato chips bag to a used toothbrush, many objects can serve their intended function only once. *Upcycling* is an emerging trend whereby one sees value in both the composition and the form of an object but not the intention. That crumpled bag that once held a few handfuls of chips can be folded into a purse or bracelet. The used toothbrush can become a pen, a doormat, or one of any number of useful objects. Although more energy is used to upcycle an object than is needed to simply reuse it, it is usually a relatively small amount.

If upcycling a particular waste product is not possible— as is the case with items like dirty diapers and cigarette butts—the product can typically be deconstructed into its component parts and used again. A used diaper or pile of cigarette butts can be shredded and separated into their respective raw materials. The resulting material, from the plastic to the organics, can be used again for different purposes. While the initial intention and form of the object is destroyed, new raw materials don't have to be extracted from the earth, and synthetic products aren't added to a landfill or some plastic island in the ocean.

It's All about the Economics

In the end all waste can be reused, upcycled, or recycled, avoiding the need to burden our planet with the constant extraction of raw materials and the introduction of synthetic ones. The challenge in all of this—whether you are trying to limit your purchasing or process waste through circular solutions—is one of economics.

In terms of waste generation, if we seriously limit our buying or exclusively buy used durable goods, we will likely negatively affect our economy, making it harder to keep our growing population gainfully employed. In terms of waste processing, circular solutions depend on waste separation, which is typically more expensive than simply burning or burying waste.

The question we must grapple with is this: Are we willing to live with moderated economic growth in exchange for a healthier planet? We can make environmental progress in the short term without sacrificing our staggering economic growth, but a long-term, sustainable solution will require fundamental changes to our culture, economy, and individual perspectives. Do we want to live in a world where we are actively destroying our planet to fuel a need to acquire physical objects? Or do we want to rethink how we create and handle our waste, making possible a more balanced—and perhaps even happier—existence?

The best part of attempting to deal with the problem of garbage is that it is something we can do immediately, as individuals. We are, after all, the root cause of garbage.

To outsmart waste, we have to understand what it is and where it comes from; then we can rethink the ways in which we create waste and what, ultimately, we can do with it.

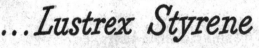

For happier picnicking Days
...Lustrex Styrene
A MONSANTO PLASTIC

Dennis Day
*star of
"A Day in the Life
of Dennis Day,"
NBC network, Saturdays
—spends a happy day pic-
nicking with wife Peggy
and son Patrick.*

Chapter 1

Where the Modern Idea of Garbage Originated

© Apic/Hulton Archive/Getty Images
© aastock/Shutterstock.com

Human refuse—"garbage"—is a modern idea that arose out of our

desire to chronically consume stuff that is made from ever more complex, man-made materials.

Consumption + Complex materials = Garbage

To outsmart waste we need to eliminate the very *idea* of waste; to do so we need to understand where the concept of waste came from and what factors brought about its existence.

Useful versus Useless Outputs

Why is it that garbage exists in the human system but not more broadly in nature? Nature is a beautiful harmony of systems whereby every system's output is a useful input for other systems. An acorn that falls from a tree is an important input for a squirrel that eats it. The by-product of that delicious meal—the squirrel's poop—is an important input for the microbes that consume it. The output of the microbes—rich humus and soil—is in turn the very material from which a new oak tree may grow. Even the carbon dioxide that the squirrel exhales is what that tree may inhale. This cycle is the fundamental reason why life has thrived on our planet for millions of years. It's like the Ouroboros—the ancient symbol depicting a serpent eating its own tail; in a way, nature truly is a constant cycle of consuming itself.

Even we humans, up until about a century ago, lived our lives in the same way: all of our outputs—from the carbon we exhaled to our feces and product waste—were cycled by nature until they became useful inputs again.

Yet today much of our waste breaks this age-old cycle by not being useful to any living organism. In the past century, the raw materials that make up our products have changed from easily degradable animal, plant, and other natural sources to highly refined, typically non-renewable resources (primarily oil). Today even when we use renewable resources (like trees), we typically render them useless outputs (like a used coffee cup) that cannot be easily recycled (due to the thin plastic lining on the inside).

This transition represents the first time in history that a species has moved away from a circular material supply chain—where every output in nature is cycled through

input   Something that is used as food or raw material for an organism. *Example:* an acorn for a squirrel.

output   Something that leaves an organism and is no longer useful to it. *Example:* one's poop.

multiple organisms until it becomes an input again—toward a linear one.

Take, for example, the plastic bag that may have been given to you when you bought this book. The useful life of a plastic bag is perhaps an hour or two—in other words, about the time it takes you to travel from the mall to your home. After that one trip, the bag typically ends up in your garbage on its way to the local landfill. Once at the landfill, it stays there, in some form, virtually forever.

Plastics, due to their molecular stability, do not easily break down into components useful to nature. Some estimates show a plastic bag taking 500 to 1,000 years to degrade.[1] We say "estimates" because Alexander Parkes invented the very idea of plastic in 1856, and not enough time has passed for any plastics to fully degrade.[2]

Additionally, a plastic bag does not just degrade like a banana peel, which is consumed by a variety of hungry microbes. Instead the plastic bag photodegrades—a process whereby the bag breaks apart into smaller and smaller pieces. The resulting particles are deadly when ingested by living things and can also contain pollutants like polychlorinated biphenyls (PCBs) and endocrine disrupters. Worse yet, they often resemble food, like zooplankton, and are inadvertently consumed by animals, such as jellyfish, who mistake the harmful materials for lunch.

product waste   An object that itself (not its packaging) becomes waste. *Example:* a toothbrush or dry pen.

But the story of garbage doesn't end with a few dead jellyfish. The United Nations Environment Programme (UNEP) estimates that there are 46,000 pieces of plastic floating in every square mile of our oceans;[3] this material, after damaging the aquatic ecosystem, may somewhat ironically end up in our order of sushi or fish-and-chips— potentially helping us get cancer earlier in life. This is just one of the many possible consequences of throwing out product waste. A recent report from a cooperative that includes UNEP and the World Health Organization (WHO) said that more than 800 man-made chemicals, including bisphenol A (BPA), were found in products we consume every day.[4]

The Birth of Widespread Chronic Consumerism

To further compound the problem of useless outputs, today we live in a world of chronic consumerism—a world where we buy much more than we need. This unique behavioral trend began just after the Second World War. In the late 1940s, our forebears had not only lived through the Great Depression but also just emerged from the biggest war the world had ever seen. Humanity needed to rebuild. Some countries started by putting returning veterans to work converting wartime factories into firms that served civilian consumption, later expanding to meet growing demand.

Due to the development and the commercial viability of plastics and other man-made materials, we were perfectly suited for the arrival of cheap, disposable products. The production of such products made it easier for common people to acquire luxuries that were once expensive or even unattainable—not to mention to buy their way out of doing dishes by hand and keep food from spoiling longer than before. Just think of the joy brought to homemakers when in 1947 Earl Silas Tupper patented Tupperware or when the scientists at DuPont invented nylon.[5]

And it worked. In the years following WWII, the economy rebounded better than anyone could have expected, and we have maintained a growing consumer appetite ever since. In parallel, and in no small part due to postwar global prosperity, the human population grew sevenfold during the same time period; a population that was just over 1 billion people at the beginning of the twentieth century is today well over 7 billion.

As you might expect from a system based largely on the production and the consumption of synthetic materials, the resulting garbage problem followed our economic growth. In 1905 product-related waste was well under 100 pounds per person per year in the United States. Additionally, it was primarily made from useful outputs like wood, cotton, and other materials that nature can use as a positive input. Since then product waste has grown by 1,400 percent.[6] What's more, 75 percent of that waste is now made of useless outputs like plastics and other complex, man-made materials that nature cannot repurpose. It is even worse when you consider how much our population has grown and that the problem grows proportionally with it.

But why should we change? The production and the consumption of cheap goods was the silver bullet that brought us out of national depression and global war and into the greatest period of economic prosperity in human history. Perhaps we should go on celebrating this achievement with more marketing and more consumption.

And it looks like we have. Many people define their lives based on their accumulated stuff. Just look at the admiration our society has for people who drive a fancy car, live in a big house, or are profiled on TV shows like *MTV Cribs* and *Lifestyles of the Rich and Famous.* Our accumulated things define our rank in society and are points in the system whereby we are "scored." I personally fall into the same quagmire: I drive a fancy sports car, recently bought a larger house, and care about how much my business grows year over year. While I'm trying very hard to change, I can tell you that it is an uphill battle.

All economic and employment growth, by definition, was and still is predicated on people's buying stuff. To the detriment of our ecosystem, there is no public policy or popular culture to curb consumerism or slow population growth. Both are direct drivers of our current definition of prosperity—a definition grounded in economic growth and measured by looking at gross domestic and national product. In fact, we seem to focus entirely on driving consumption instead of curbing it.

Recently, I was having a conversation with the chief executive officer of a major North American waste management company, who told me that the size of his business is directly related to the size of the economy (typically with a lag of about a year, as it does take time for the objects we buy to become waste). With the modern market economy, we value companies not only on their objective size but also on their growth and potential for more growth. Growth is our economy's insatiable goal, and the larger our economy, the larger the waste problem.

The Natural Controls of Chronic Consumption

Where we are today is entirely natural, and, in a way, it is built into our DNA, as the objective goal of all living organisms is to live and grow, both as individuals and as a species. These innate desires—some of the fundamental cornerstones of sustaining life—are met by gathering food, building a home, and reproducing.

Have you ever been stuffed after a big dinner only to find yourself downing a large piece of cheesecake that could have served as a meal in and of itself? In all living things, as in humans, the desire to consume is largely uncontrolled. If you put a big pile of sugar in front of a healthy mouse, it will eat well beyond what it needs to survive each day until it becomes obese and diabetic and eventually dies. The poor (yet sugar-rich) mouse simply cannot control itself from binging on easily attainable calories.

The reason why nature has sustained itself in relative harmony for millions of years is because a number of external factors control the ability (or inability) of that mouse to gain food, find shelter, and have offspring. Instead of a mouse's controlling its individual consumption, its ability to consume is indirectly controlled by other organisms and nature at large. Even if the mouse wants infinite sugar, in nature it has to contend with predators, competition for sustenance, and a general lack of abundance.

Predators   In nature our friend, the soon-to-be-obese mouse, has to worry about a whole host of other animals, from the friendly barn owl to the sly house cat—creatures that would love to make the mouse their supper. Humans have solved that problem by controlling, avoiding, or killing all of our natural predators and by preventing and combating disease (perhaps our last true predator). These days we don't see wolves and bears roaming around our villages like they used to, and when we get sick we have better treatment than ever before.

Competition for sustenance   Anytime there is food
in nature, the "dinner bell" is heard (or smelled) by every
potential diner in earshot (or perhaps "noseshot"), making
it hard to gorge oneself into obesity. If you leave a pile of
nuts and berries in the forest, within minutes a whole host
of woodland creatures would get in on the party. The pile
would be gone by the time late-comers arrived.

Because we effectively match the supply of goods with
demand (and in most cases overproduce), humans simply
don't have competition for consumption in the way there is
competition in nature. So when that new iPhone sells out,
we simply make more (God forbid limiting supply!). It is bad
business to run out of stock.

Lack of abundance   In nature it takes time and hard
work to find food. Just think about the energy you would
expend to feed yourself if you were dropped in a forest
while reading this book. First, I would apologize to you for
the teleporting powers of this text. It would be very hard for
you to find food, and you may not come out alive (unless
you have trained or watched enough of those TV survival
shows). Either way, the slimy grubs and the bark you would
eventually have to consume wouldn't be quite the same as
ordering Chinese takeout, and they would be much harder
to find than the delivery guy on his Vespa outside your door.

In the human system, with the amazing progress of science,
we produce more food than we can possibly eat and more

products than we can possibly enjoy. According to a recent study, 30 to 50 percent of the food we produce is thrown away due to its appearance or a lack of demand.[7] That's 1.2 billion to 2 billion tons of food per year being thrown out![8] While that may seem staggering, we often throw away durable goods, such as clothing, before we have even used them, just because our tastes have evolved since we bought them.

No predators, no competition, and uncontrolled abundance put us in a truly unique position. With few external controls to speak of, we are that gluttonous mouse, gorging itself on the proverbial pile of sugar.

But it doesn't have to be this way. By rethinking how we produce and consume, we can live sustainably and return to a harmonious relationship with nature. As individuals we can impose on ourselves the same limitations that predators, competition, and lack of abundance have placed on our nonhuman earthly co-inhabitants. From there a personal shift in consumption habits can move outward through our friends and networks, ultimately affecting the larger society in the form of our culture and perhaps later our laws.

So far as ideas for where to start are concerned, nature itself seems to have some pretty good ones.

The Wisdom in Mimicking Nature

Nature simply has no concept of garbage, or useless outputs. Think about when your dog eats a plastic object, thinking it is food, or your cat chews on an extension cord. In nature all outputs are useful. It is a natural wisdom that we should echo, not in the innocent ignorance that leads some unwitting creatures to eat inedible trash but in its fundamentality.

The emerging field of biomimicry, championed by luminaries from Janine Benyus to Paul Hawken, has effectively commercialized this exact notion. They have found repeatedly that taking inspiration from nature can help us solve concrete human engineering challenges.

For example, chemical companies seeking to develop self-cleaning paint turned to the lotus—a plant that needs to keep the surfaces of its leaves clean despite living in muddy ponds and swamps (which are, admittedly, strange environments for something so beautiful and seemingly pristine).[9] To help stay dirt-free, the lotus plant evolved tiny ridges and bumps that stop water droplets from spreading across the entire surface of its leaves. Water beads form, slide down the leaves of the plant, and carry off dirt with them. Taking a tip from nature, paint developers created paint that leaves tiny bumps when it dries—helping water form droplets to carry dirt away.

Another example of effective biomimicry is the Japanese Shinkansen "Bullet Train." At more than 200 miles per hour, the Shinkansen is the fastest train in the world. Due to changes in air pressure and the speed of the train, every time the train came out of a tunnel it would create a micro sonic boom. The Bullet Train became something of a noise problem, with villages miles away complaining. Eiji Nakatsu, avid bird-watcher and the chief engineer on the case, found that modeling the front end of the train after the beak of the kingfisher—a bird that can dive into bodies of water with almost no splash—not only solved the sound problem but also saved 15 percent in power while increasing the train's speed by 10 percent.[10]

When it comes to mimicking nature to our own benefit, we should be aware that nature doesn't innovate at the rate humans do.

To learn from nature and be responsible to it in the long run, we should focus on creating useful outputs rather than outputs that are useless (and potentially toxic). It is a proposition where all stakeholders are winners.

As individual consumers, we should look to buy products that are made from natural materials (their by-products make for useful outputs), ideally avoiding complex materials altogether. By consciously controlling our consumption and buying products that produce useful outputs instead of useless ones, we can take a big step toward eliminating the idea of garbage.

Chapter 2

The Role of the Individual Purchase

When looking at the root cause of garbage, we as consumers bear a large part of the responsibility. Garbage is predicated on our individual consumption. If we don't buy something, it can never become garbage.

The manufacturers that make our products are here to serve the desires of consumers (you and me). Sure, by marketing to our desires they may influence what we want—or even introduce things we never knew we wanted—but in the end we are the individuals who pull the trigger and trade our money for those goods. No one is forcing us to buy anything. By contrast we voluntarily, and in fact willingly, buy things on a daily basis. We even gain pleasure in the act of buying. Consider the recent emergence of "retail therapy," a pop-culture concept promoting the act of shopping as a way to beat depression.

Consumerism, in a way, is something of an addiction. We almost *need* to consume; we are constantly chasing after the next new *thing* (or *high,* for the sake of this metaphor), and our appetite to consume is never satisfied. One key difference between global consumerism and individual addiction is that this destructive habit doesn't just harm our individual bodies; it affects our planet in a real way.

Our Most Powerful Vote

Consumer product companies spend millions of dollars every year on market research. The purpose of their existence is simple: to make a profit by figuring out what we want and putting that in front of us so that we can buy it. To be fair, companies also use advertising to influence our purchasing choices by exploiting our insecurities. They don't just cater to our wants; they also help form our desires. As powerful as marketing can be, we are still ultimately in control of what we purchase.

When we buy something, we are actively voting for more of a particular product to be made. On the other hand, when we don't buy something, we are voting for less of that good to be produced, and the vote we cast is exceptionally powerful. Just imagine if we all stopped buying chewing gum altogether for just a month or two. First, retailers that sell gum would stop carrying it—they are in the business of carrying only what sells. With no retailers displaying their wares, it wouldn't take long before the entire gum industry collapsed on itself. Perhaps we'd have worse breath as a result, but we would no longer see the ugly black dots that littered chewing gum leaves on sidewalks.

It's true—as a result of the gum industry collapse, thousands of employees would find themselves unemployed. But this quandary begs the most important questions in all of this. What is more important: the growth of our economy or a sustainable existence? How many humans should the planet support? Is more always better?

The consumer vote, unlike the votes we cast in elections, is one that you make daily with your hard-earned dollars. It is profound, and it fundamentally controls what is made and what is not—what ends up as garbage, and what doesn't.

We Voted for Disposability

A century ago consumption didn't exist in the way it does today. People worked long hours (they also "walked uphill on the way to school and uphill on the way back") and had less time to go shopping. In fact, going shopping often meant heading to Main Street and picking up the necessities from the few local stores that offered them. Products were generally more expensive, and most people bought only what they needed. Shopping for the sake of shopping didn't really exist, and consumers prized durability. When shoppers bought an object (except consumables like food), they expected it to last; when it did eventually break, rather than throw it away they would fix it. If overalls tore, their owner would patch the tear. They would cobble their shoes when the soles wore out and fix kitchen tables when they broke.

Then along came Henry Ford and the perfection of mass production, ushering in an era of mass consumption. Today when we buy a table, we expect it to break within a handful of years. If our pants tear or our shoe soles wear out, we choose to throw them out and buy new ones. Even before a hole appears or a sole needs replacing, our ever-changing fashion aesthetics often render those jeans or shoes as waste with the next season. Just imagine the reduction of your clothing budget if the idea of fashion evolved at a slower pace or not at all.

To serve their shareholders, manufacturers seem to make sure that objects become physically outdated, or outdated in the eyes of consumers, as soon as possible; if something breaks easily or if we think the next version of an object is better and cooler than the one before, we're more apt to discard the old in favor of the new. Planned obsolescence is simply good for business.

Planned Obsolescence

The concept of planned obsolescence originally emerged as an attempt to turn the economy around during the Great Depression. In 1932 Bernard London published a book called *Ending the Depression through Planned Obsolescence* in which he suggested that governments should impose legal lifespans on products, after which consumers would be required to return the object to be destroyed.[1] This would spur more consumerism and help put everyone back to work.

Brooks Stevens, an American industrial designer, helped popularize the idea of planned obsolescence in the 1950s. He defined the concept as "instilling in the buyer the desire to own something a little newer, a little better, a little sooner than is necessary."[2] In his mind it wasn't as much about making objects that break or requiring an object's

> **planned obsolescence**   A policy of planning or
>   designing a product with a limited useful
>   life so that it will become obsolete, that
>   is, unfashionable or no longer functional,
>   after a certain period of time.

destruction after a certain period of time but rather making
the consumer always desire something new.

Although planned obsolescence never became law, it was
widely adopted by business in principle and is still in action
today. There are two categories of planned obsolescence:
obsolescence of desirability and obsolescence of function.
In lay terms, manufacturers can render an object waste
before it breaks through either fashion or planned
breakability.

Fashion   Would you wear clothing with shoulder pads?
Sport a fanny pack or tracksuit? Put a scrunchie in your hair?
If the styles of the 1980s were still in fashion, you might, and
all of those items would be flying off the shelves. Fashion is
all in our heads. When *Vogue* and *ELLE* magazines say that
flat shoes are in style, the heels hit the trash. When someone
decides that heels are back in, the flats are out. Though it
seems unnecessarily wasteful when scrutinized, fashion is
good for business; it renders an object as waste well before
it breaks and in turn drives incremental consumption.

Planned breakability   Perhaps the most famous case of planned breakability is that of the light bulb. Today, more than 150 years after the light bulb was invented, an average modern incandescent bulb lasts 750 hours.[3] So why is there an incandescent light bulb in Livermore, California, that has been burning nonstop for 110 years?

It's because in 1924 in Geneva, all the light bulb manufacturers got together and created a cartel, at the time called Phoebus, to cut the life of a light bulb. By making the filaments more fragile than they needed to be, light bulbs would burn out quicker, and the cartel could sell more.[4] It was so serious that manufacturers were fined if their light bulbs lasted more than 1,500 hours. Before 1924 the average bulb lasted about 2,500 hours. Within a decade it was at 1,500 hours and has since declined to today's 750.[5]

The same thing happened when nylon came to be. The first nylon prototypes from DuPont were so strong they could tow a car. But durable stockings don't make for frequent purchases, so scientists created a more fragile material.

Today things aren't much different. Besides light bulbs and nylons, many inkjet printers contain a counter chip that will effectively shut off the printer once it has printed approximately 20,000 pages—even if nothing is wrong with it. And have you ever tried changing the battery on an iPod, iPhone, or iPad? In 2004 and 2005, class action lawsuits were brought against Apple regarding diminishing battery life.[6] Before then company policy was to instruct users of iPods with defunct batteries to buy the new model. As discovered during the lawsuit, Apple consciously designed the battery to fail after a year or two.

Planned obsolescence is perhaps the biggest injustice to consumers committed by the manufacturing industry. It is one thing for a consumer to be manipulated by changing fashion, but it is quite another to sell objects that are designed to break.

We Are in the Service of Consumerism

Any economist will agree that consumerism is essential for capitalism to survive. For our standard of living to increase, not only must consumers buy but they must buy more every year. Without perpetual and growing consumption, the economy declines, and decline begets more decline. The conventional sign of a healthy national economy is, after all, measured by gross national product, a measure of the quantity of goods and services people consume.

So are we all in the service of consumerism? Today I would argue yes. Most jobs in the world serve to facilitate consumerism. From factory workers in China and sales associates at your local department store, to most lawyers, accountants, and real estate agents, the list of professions that facilitate consumerism in some way is a long one. Like our friend the diabetic mouse, gorging himself on that pile of sugar, humans can't help but consume, and there are many repercussions of growing consumerism. From health problems like obesity and diabetes to environmental problems like global warming and growing landfills, rising consumerism leaves an increasingly large mark on us and on our planet.

But consumerism isn't all bad. It drives innovation, which typically increases our standard of living. Just think of how many people have access to a flushing toilet and a mobile phone today. Now ubiquitous in many nations, even the wealthiest people could only dream about these luxuries a century ago.

So the answer to outsmarting waste must contain within it a serious conversation about controlling our consumerism. Over the long term, we need to ask the question of whether driving happiness through consumerism is sustainable. In the short term and as individuals, we can make important changes by consuming consciously.

Buy Stuff, but Buy It Differently

While we may be in the service of consumerism, we are not slaves to it. We are serving the idea of consumerism voluntarily and happily. In fact, sometimes we even spend beyond our resources and go into debt to gain the positive feeling of buying something.

If we all changed our daily vote (the stuff we buy), within a very short time we could solve the global garbage crisis. We would probably have more time on our hands and be happier overall too; the constant hamster wheel of working to have happiness-giving products to consume would be a thing of the past.

In the absence of unanimous global rejection of consumerism, we can still reduce our impact on the planet as individuals. For the solution of changing our buying habits to be accepted by popular culture in the short term, it needs to come without harming our economy as we traditionally define it. Only then can we move toward a culture that will accept a form of economy not driven solely by size and growth.

Thrift—or moving back in time to the village mentality—is not the short-term answer. Experiments to take this vision mainstream have failed on all accounts. The economic paradox of thrift states that if everyone tries to save, aggregate demand will fall; this will in turn lower total savings in the population because of the overall decrease in consumption and economic growth. In other words, the concept of buying stuff is central to our current view of economic progress.

To shift our culture of consumption in a sustainable way, we first need to become conscious of what we buy.

Buy Consciously

Just think of the horror any decent person would feel if we were hiking together in the woods and I dropped the plastic wrapper from an energy bar I was eating on the path behind me. The reaction would be quite different if I finished eating an apple and tossed the core. Rather than horror my companion might even react with a smile, knowing that I was contributing a useful output to the forest through which we were hiking.

Without exception, everything we consume will one day become waste. If you eat a pizza, you will eventually poop it out. If you buy a TV, one day it will break and almost certainly end up in a landfill. Your clothing will tear or go out of fashion; even you and I will one day die, and our bodies will become waste.

But not all waste is created equal. The pizza, our bodies, and that apple core I might have tossed in the woods— all will become a useful output: organic waste, which is an input that little organisms and other grubs will happily eat. Perhaps the best way to go when we die is to be composted—turned back into food for deserving organisms—rather than burned or buried. If we do, our bodies will become beautiful plants much faster than they would if we slowly mummified in coffins deep underground. It also hedges against our becoming a zombie and eating our friends' brains for supper.

The TV and the clothing (if made from synthetic material like nylon), however, will become useless outputs. The TV, especially if it's a tube TV, will turn into a toxic mess that will poison everything near it. The clothing, while less toxic than

> **compost**   A mixture of various decayed
> organic substances that can be used to
> grow plants. Looks and smells like soil.

a TV, will photodegrade into smaller and smaller poisonous molecules. Simply put, anything that is made from synthetic or complex materials will result in useless outputs. As such, when you buy stuff, first consider what will happen to it when it inevitably turns to waste—and choose products that become useful outputs.

Making this sort of lifestyle change is easier than it may seem. You can start by buying stuff with minimal packaging, or no packaging at all, and trying to avoid products made from synthetic materials. When buying a chair, buy a wooden one instead of a metal one with foam stuffing and nylon covering. When buying food, choose fresh fruits and vegetables (without using those plastic bags) instead of packaged ones. You may find yourself getting healthier and creating fewer useless outputs at the same time.

Buy Durable

When my family (well, really my parents with me in tow) was escaping communist Hungary in 1986, my dad pawned his very hard-earned gold watch, allowing us to have enough money to bribe the border guards to let us out. If he had had a cheap, plastic watch, I would still be living in Budapest, and you probably wouldn't be reading this book.

Imagine if we valued durability like we did in the past. Because major consumer product companies are in the service of what the majority wants, they would shift from making high volumes of disposable, low-cost products to lower volumes of durable, high-cost products. With no market for cheaply made goods that break easily, the idea of planned obsolescence could potentially come to an end—at least the kind based on the lifespan of products rather than fashion. We would also consume less—purchasing durable products would be more costly—but the economy would not suffer because we would probably still spend the same amount of money.

But seeing value in higher-quality goods would not just mean a world in which light bulbs lasted longer. Buying durable would be a sign that we have really shifted our outlook to long-term thinking; in a way, we would acknowledge that our decisions have long-term impacts. We would be making fewer but more-important buying decisions. Because our workforce would be making more-complex, higher-end objects, we would need better education, and our standard of living would increase, perhaps even faster than it does today. The epidemic of morbid obesity would decline, as people would eat less but better food. Birth rates would decline, as it would be more expensive to have lots and lots of kids.

All we have to do is change our purchasing preferences from disposable to durable goods. The amazing thing about capitalism is that it is in the service of our wants and needs, not the other way around. If we change our purchasing habits, manufacturers will change their products and services. The more we spend, the more power we have to change what the market produces. It's been said that "the customer is always right." While there is some truth to that statement as it is, when it comes to shifting market priorities we can confidently add, "but the more they spend, the more right they are."

Practically speaking, prioritizing durability and quality over affordability in the products we buy may mean that we can't buy everything we desire right this second; we may need to save up our money to buy that beautiful, durable pen instead of a cheap, disposable one. Even though the ability to own that pen is a bit farther off, the experience of writing with a durable pen may be better than with the disposable one, and you may never have to replace it. You may even be able to give it to your children, and they to theirs.

The concept of prioritizing the purchase of durable goods over disposable ones can be applied to all products except

durable good   A good that lasts for multiple uses and typically can be repaired when broken. *Example:* a pair of eyeglasses.

consumables like food and cleaning products. When you buy clothing, consider buying well-made, timeless pieces that will last and stay in fashion instead of cheap garments that you may not ever wear. In the end this change in shopping behavior and the resulting reduction in useless outputs may increase overall well-being and happiness of both you and our planet.

Buy Used

Buying new durable goods has a lower environmental impact than buying disposable goods, but it is possible to take this type of thinking even further. Buying used durable goods, instead of new, saves a perfectly decent product that one person no longer needs from actually becoming garbage. It also prevents the need to make a new one. The good news is that used objects are typically durable; if they weren't, they wouldn't have lasted long enough to show up on the secondhand market.

Buying used is important because it avoids the need to manufacture a new object, making the environmental benefits of buying used immense compared with any other form of purchasing. You may find the secondhand market— garage sales, antique dealers, thrift stores, and websites like eBay and Craigslist—to have many of the durable goods you might otherwise purchase new. The volume of used stuff out there is amazingly large—partly because we seem so wired to prefer new products—and there is a direct economic benefit to the consumer because goods on the secondhand market tend to cost less than they would new. So maybe the next time you want to buy a lamp, instead of buying a new, durable lamp, you could find that same amazing lamp on the secondhand market and cut having to produce a new lamp out of the picture.

Because the concept of buying durable goes hand in hand with the concept of buying used, it can be applied only to non-consumables (you can't, and wouldn't want to, buy a used pizza). When you buy clothing, consider going to a vintage clothing store instead of a department store or designer. Fashion cycles anyway, so wouldn't it be more authentic to wear something actually from the 1960s or 1970s instead of something made today to mimic the style of a bygone era? When you buy furniture, why buy new stuff with paint-on scratches when you can get an antique with real ones that has already proven itself to last for decades?

It is right around this level of change—when new objects are not created—that our economy may start to suffer by current measurements of economic health. Even though the vintage shop or online retailer will be making money, manufacturers no longer will because no new product is made.

The final step in this progression from hyperconsumption to sustainable consumption is to reconsider the act of purchasing altogether.

Stop Buying

Have you ever walked into a garage, attic, or storage unit to see it filled to the brim with junk that isn't being used or in some cases was never even opened? Perhaps the ultimate irony of our chronic consumerism is the fact that we often accumulate stuff for the sake of buying it, and we spend money storing it only to spend more money when we throw it all out in the end. "Purposeless consumerism"—which, in a way, isn't much different than paying for the ability to take an unused good from a store and throw it out—is really the worst kind of consumerism.

The concept of "degrowth" asks: do I really need this object, or am I buying it because I like the feeling of buying something?" It originated from the ideas of ecological economics and anti-consumerism.[7] The key to the concept is that reducing your consumption will not reduce your well-being; rather, it will maximize your happiness by allowing

Buy Differently to Make an Impact

- Stop buying
- Buy used goods
- Buy durable goods
- Buy consciously

you to have more time and savings to spend on things like art, music, family, and community. Today we consume 26 times more stuff than we did 60 years ago. But ask yourself: *are we 26 times happier?*

Before you call me "comrade" and disregard the point, consider what Mahatma Gandhi said: "Not all our gold and jewelry could satisfy our hunger and quench our thirst."[8]

Buying Our Way out of the Garbage Problem

The modern idea of garbage comes from the confluence of synthetic materials (which make for useless outputs) and chronic consumerism. To solve garbage we first must embrace the fact that we vote for it every day by buying stuff—especially disposable stuff.

Let's start outsmarting waste by being conscious consumers. Let's start buying stuff that results in useful outputs rather than useless ones—and buy only what we really need. When buying consumables like food, avoid packaging; it makes up a tremendous volume of our daily waste, and fresh food is better for you anyway. When buying non-consumables like a mirror or lamp, try to find something durable; and if you have to buy something durable, buy something used rather than new. And before you buy anything, ask yourself: *do I even need this?*

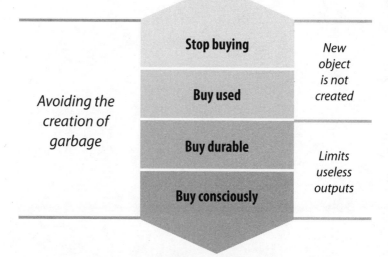

Be conscious when you buy. If you aren't, it's like going into a voting booth and randomly picking who will be the next president. Every dollar you spend is the most powerful vote you will ever cast.

Chapter 3

Our Primary Global Solution to Waste: Bury It

W hen I was a child, I had a pet rabbit that lived in a large cage on our apartment balcony. Every day I would feed her the vegetable peelings from our kitchen; she would happily eat them, later pooping out whatever her body didn't use as spherical, pearl-like droppings in one corner of her cage. She would spend the rest of her time hanging out, dreaming perhaps about nice boy rabbits, in another corner of the cage. I never once saw her venture near the "poop corner" unless she had some specific business to do. Come to think of it, if I were that rabbit, I probably wouldn't either.

The desire to be as far away from one's own waste as possible seems to be hardwired in us. Landfills constantly face NIMBY ("not in my backyard") challenges when getting zoned, and property values are lower near sewage treatment facilities, landfills, and composting sites. People simply don't like hanging out near waste. Perhaps that is one of the reasons why we invented the toilet. If you deconstruct what a toilet is, beyond being a nice ceramic seat, it's a device whose purpose is to move our waste far away from us as fast as mechanically possible.

The rate of innovation in the waste management industry is much lower than in other industries like beverages or pharmaceuticals. It's simply not desirable to hang out in garbage; or, by extension, it's not sexy to work in the field of waste management. When you ask a child what he wants to be when he grows up, you expect to hear things like "policeman" or "baseball player" but never "garbage man." And if a child did express an interest in eventual employment in the waste management industry, you'd probably cringe and explain that being a doctor or lawyer might be a better path.

A key aspect of outsmarting waste is rethinking our relationship with it. Instead of running away from waste and hoping something else will take care of it, we must actively take ownership of it. We should not only own the problem but embrace it. While I still advise being repulsed by one's own feces, if we make solving garbage sexy and celebrate it, we can spur much needed innovation in the field.

The Birth of the Dump

Ever since garbage became a growing problem about a century ago, we have used one primary solution to deal with it: bury it. Even at the beginning of twentieth century (and naturally before), we simply dumped garbage in the most convenient locations—from oceans to wetlands to any given "wasteland." It took another 25 years for the United States to set up any regulations about where we could pile our garbage; and by 1934 the US Supreme Court had outlawed the dumping of waste into the oceans (which at that time was incredibly common).[1] Today 25 percent of all waste still ends up in our oceans, even though it is illegal to dump it there.[2] Our waste disposal "solutions" thus far have resulted in the birth of the Great Pacific garbage patch—a pile of degraded plastic sludge 10 meters deep and as large as the state of Texas.[3] While the Pacific trash vortex is perhaps the most notable, there are currently five ocean gyres—each its own giant aquatic garbage patch.

The only real differences between our garbage disposal methods at the beginning of the twentieth century and today are that our burying techniques have become more sophisticated and regulations have stopped us from dumping waste directly into environmentally sensitive areas. Instead of just putting waste into a big pile, we now dig a gigantic hole and line it with thick plastic to reduce the risk of toxic leachate runoff ("landfill juice"). After a day of filling the dump, we cover it with a thick layer of soil to keep away the vermin and keep the smell in. We even have an entire gas piping system built into landfills to get out the methane—a highly combustible natural gas that is 20 times worse for the environment than carbon dioxide. Methane is also the stuff our farts are made of, and it is sometimes converted into energy at modern landfills.

The Economics of Waste

Burying garbage is our most common global solution because of the economics of our waste. A simple explanation of supply and demand is that as demand for a product increases, supply decreases (as people are buying more) and prices go up. On the contrary, as demand for a product decreases, supply increases (as people are buying

Supply and Demand for a Normal Product (e.g., a Potato)

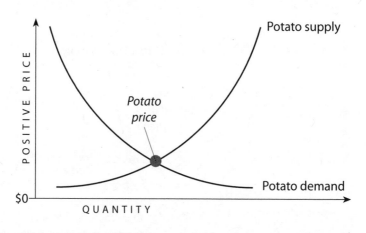

less) and prices go down. Basically, the more supply there is, the lower the price will be; the more demand there is, the higher the price will be.

Strangely, all supply-and-demand curves found in economics textbooks show the intersection of the supply-and-demand lines in the positive quadrant—implying that the price of a commodity will always be greater than $0. It makes sense: no matter how many potatoes are being produced, we could never imagine the potato farmer paying you to take his crop. The price may be exceptionally low, but it would never be negative. That would be like paying your employer for the privilege of working at your job or your teacher paying you for the privilege of teaching you.

Garbage, on the other hand, might be the only commodity to break this rule, as it is the only commodity with negative

Supply and Demand for Garbage (e.g., Dirty Diapers)

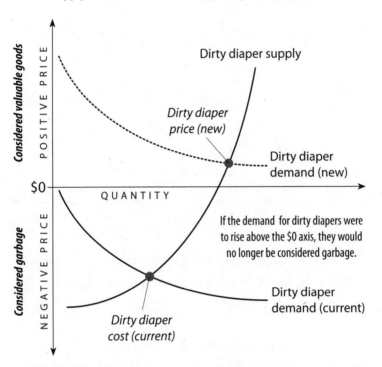

demand. In other words, we actively pay (our waste management bills or our taxes) to have our waste taken away from us. As we have more waste (more supply), the price for getting rid of it gets more expensive. In other words, the negative value of garbage becomes increasingly negative, meaning that our cost to dispose of it goes higher and higher.

But there is something really special in all of this. Imagine that you were going to build a house and the hardware store paid you to take building materials or that you went into a restaurant and the waiter paid you to eat your meal. It would be amazing! Leveraging waste as a raw material is kind of like that. If you view as a valuable material something that others view as waste, your raw material costs to use that material would be negative. This would change only when others also start sharing your perspective and the demand for the material increases. If that were to happen, the particular material stream that we used to know as "garbage" would no longer be considered garbage. Instead it would be just another material with a positive price.

Garbage is something we are willing to pay to get rid of. As such, if we were to start seeing it as valuable and convince others to do the same, the cost of that material would rise (in other words, the cost to dispose of it would decrease) until it crossed the magical $0 line—at which point it would no longer be economically defined as garbage.

The Problem of Mixed Garbage

We bury garbage for two reasons. First, because garbage is seen as a commodity with negative value, the best way to make money on such a commodity is to dispose of it as cheaply as possible. There is nothing cheaper than doing nothing and, in the world of waste management, putting something in a pile is the closest thing there is to doing nothing.

The second reason is that human garbage is much more complex than the waste of any animal. Animals, such as a bird, have three categories of waste: their excrement, their body parts (and in the end their body), and what they exhale. For humans it is not just our nail clippings and poop; from our cigarette butts to dirty diapers and everything in between, we create an amazing mixture of useless outputs. The range and the number of materials are as infinite as the range and the number of products we have the option to buy. And because our garbage is mixed together into one big, inglorious pile, it is both difficult and costly to separate.

Imagine if you walked into a supermarket, took every product off the shelf and out of its packaging, and mixed it all into a pile in the middle of the floor (and then promptly left before someone caught you). The owners of the supermarket would be furious—and not just because you made a mess. It wouldn't make any financial sense to separate everything and put it all back into its appropriate packaging. Instead the owners would likely throw everything out and buy new products to restock their depleted shelves.

Similarly, waste management companies are left with the challenge of how to solve, as cheaply as possible (while complying with local laws and regulations), the problem of a highly complex material that is basically infinite in variation. Separating it out into what it's made of would allow the material to be carefully processed, but that is too costly, and as such burying remains the logical approach.

But what is the end game? We live in a world of finite raw materials, a world in which, in the 1970s, we started taking out more from the planet than it could replenish. If we keep burying our waste, one day we will either end up on a planet made of landfills (remember the Pixar movie *WALL-E?*) or run out of raw materials. Neither outcome is desirable, and the two don't have to be mutually exclusive.

The Irony of Turning Useful Outputs into Useless Outputs

One of the great ironies of modern-day landfills is that they render what could easily be useful outputs—organic waste like those kitchen scraps my rabbit loved to eat—into useless outputs. Because landfills don't allow oxygen throughput, any organic material that is dumped (up to 70 percent of all material sent to landfills) will not decompose.[4] The three necessary components for decomposition—sunlight, moisture, and oxygen—all are hard to come by in a landfill.[5]

The stuff we bury is far more likely to mummify than to break down. In fact, when some old landfills have been excavated, 40-year-old newspapers have been found with easily readable print.[6] Researchers at the University of Arizona even found still-recognizable 25-year-old hot dogs

and heads of lettuce in landfills across the country.[7] Those
very hot dogs and newspapers could have become useful
outputs if they were composted properly. In fact, those
materials probably would have already cycled a few dozen
times through various organisms if not for being suffocated
in a landfill. Instead they did little other than take up space,
trapped in a man-made garbage pile.

Because landfills are so tightly packed, any degradation that
does occur is done anaerobically, or without air. Anaerobic
processes generate tremendous amounts of methane
gas—a major contributor to global warming—and can even
turn the landfill into a bomb.

Exploding landfills are no joke, and they occur frequently—
even in the United States, where you'd think landfills would
be highly regulated and well constructed. In 1995 a landfill
exploded in North Hempstead, New York, destroying a
driving-range snack bar.[8] In 2000 a house in Rochester
Hills, Michigan, exploded in the middle of the night when
landfill gases migrated from the nearby Six Star Landfill.[9]
The residents were able to escape, but their dog died in
the blast. These are only a few of the countless examples of
landfill fires and explosions, but we still keep burying our
trash and hoping for the best.

So be conscious when you throw something out; anything
you put into your garbage is destined to live forever in a
smelly, potentially explosive mountain of waste.

Waste Pickers

While it is illegal in most developed countries to enter a landfill and sort out valuable materials to sell, it is quite a different scenario in emerging countries such as Brazil and India. In these countries there are millions of *cartoneros,* or waste pickers, who sort through landfills and other garbage locales to root out valuable materials.

In countries like Brazil, both the government and the public have embraced this informal economic sector and created prudent regulation and strong support systems. The story is quite different in other countries, such as Argentina, however. After the financial crisis hit Argentina in 2001, many able-bodied workers lost their jobs and hit the streets of Buenos Aires, scavenging for sellable garbage to support their families. The influx of *cartoneros* was seen as leading to an increase in social problems, such as police corruption (forcing the *cartoneros* to pay bribes to the police for each cart they use) and frequent fights between residents and *cartoneros* over messes left after garbage sorting in the middle of the night.

While it is very hard to know how many waste pickers there are globally, the World Bank estimated in 1988 that 1 to 2 percent of the global population earns a living through waste picking.[10] A 2010 study looking only at India put the local population of waste pickers at a staggering 1.5 million people.[11]

As with all waste solutions, waste pickers are inextricably linked to the economics of garbage. In emerging markets where the cost of living is relatively low and people are willing to work for less income, it is economically viable to have a vocation on a landfill, searching for aluminum cans and soda bottles to be sold to recycling centers, or cooperatives. The expected wage for a waste picker ranges from $1 per day in Cambodia[12] to $100 per day in Belgrade.[13]

Even in developing markets where there are large populations of waste pickers, the local landfills are growing at staggering rates. This begs the question: What is the future of our global landfills?

The Future of Our Landfills

While a lot of landfills still opt to do as little as possible, some landfills have started to show that there is value in waste. Many landfills separate out yard waste and other bulky organic material before the garbage is dumped; it can then be composted instead of landfilled, rendering that organic material a useful instead of a useless output.[14]

While most modern landfills try to collect and burn off—flare—the methane that they produce (primarily to avoid explosions), a few of these landfills (less than 10 percent[15]) also capture the energy from the burning process, turning the landfill into something of a power plant.[16] These positive steps show that there is value inherent in our garbage.

Perhaps one day, likely when the costs of oil and other raw materials are high enough, we will start to view our landfills more as repositories of valuable materials that can be mined, separating out materials that may be used again.

Just as the first step in outsmarting waste is to be conscious consumers, we need to be conscious that putting our outputs into our garbage cans is probably the worst place to put them. Anything that goes in the garbage will be rendered into a useless output—even if it could have easily been useful elsewhere.

Chapter 4

The Energy Inherent in Our Waste

From newspapers to hot dogs, all objects have an inherent amount of energy—their "caloric value." Simply put, *caloric value* is the amount of energy that is released when a particular thing is burned. Some objects burn at a positive caloric value, including candles, cigarettes, or basically anything that will continue to burn after you put a lighter to it to get it going. This can easily be calculated in a laboratory by measuring the amount of heat that the object releases per gram and subtracting the amount of energy that was used to get the burn going. Objects with a negative caloric value, on the other hand, consume more energy than they produce in the process of burning.

Calories from items with a positive caloric value are exactly the same type that we try to avoid when we go on our annual New Year's diet. In other words, if you took sugary, buttery, oil-drenched, icing- and sprinkle-topped doughnuts (yum), they would burn much better (giving you more calories) than things like asparagus, celery, apples, and other foods with a negative caloric value.

So what does this have to do with garbage, you ask?

caloric value   The energy released by an object when it is burned or digested.

The Birth of Burning

Folks in our society who are charged with garbage disposal sometimes have trouble finding a nearby dump. Sometimes landfills are hard to build nearby; perhaps angry citizens don't want one to be built near their homes, or the geography of a particular area is not conducive to building one (the land could be too valuable or simply too hilly for a landfill to be built). Without a local dump, the challenge of waste disposal typically becomes largely one of cost. It is not cheap to transport waste, and the farther away a landfill is from where the garbage is generated, the more expensive it becomes to dispose of the waste.

Burning waste originally came about as a way to solve the problems of both cost and space—incineration facilities don't require a lot of room, and they can process a large amount of waste per day. Even though much of our waste burns at a negative caloric value, enough burns at a positive value (primarily plastics) to make it relatively cheap to burn waste.

The first incinerator in the United States was built on Governors Island in New York just before the beginning of the twentieth century.[1] Within half a century, there were hundreds of incinerators in operation across the United States. None of these incinerators actually produced any energy—they simply got rid of our garbage. Today less than 10 percent of global incinerators produce energy.[2]

The Environmental Impacts of Burning

If you burn waste yourself in your backyard, the environmental repercussions are enormous. Because you would be burning your waste at relatively low temperatures (around 1,000 degrees C, or 1,832 degrees F), the backyard burning process produces a whole range of toxic compounds. Nitrogen oxides—a group of compounds that are responsible for smog, acid rain, global warming, and ozone depletion—and possibly a whole range of other toxic materials and volatile organic compounds are released into the atmosphere when you burn trash at home.[3] Backyard burning is also frequently the cause of residential and forest fires. In Minnesota, for example, the uncontrolled burning of garbage, brush, and grass is responsible for 35 percent of wildfires.[4]

To be safer, incineration facilities are designed to burn waste at very high temperatures, sometimes even using plasma torches that are six times hotter than your backyard fire. Because the hotter a flame is, the more it will break down molecules into simpler and simpler units, incineration reduces—but does not eliminate—the negative outputs associated with burning garbage. These sophisticated incineration facilities typically leave us with ash that must be disposed of in designated toxic waste landfills, as well as dioxin, furan emissions, and various levels of heavy metals.

The tragedy of all of this is that the vast majority of incineration facilities don't produce any energy at all; the amount of energy they could recover from the burning process is relatively small and many times not worth the investment. They simply take waste material like plastic— material that took a tremendous amount of energy to produce—and turn it into ash, carbon, and other toxic outputs. The key benefit of burning is that the waste, in its original form, does not have to end up in a landfill.

Even if some incinerators do produce energy, what sense is there in turning plastic into the lowest form of useful output possible? By burning a yogurt cup, for example, incinerators essentially destroy all of the energy it took to extract and convert oil into a refined petrochemical and then turn that chemical into a useful container. Instead, why not melt that discarded plastic cup into a new plastic fork? In the process, we would be eliminating the need for new plastic— which is quite environmentally costly—to be harvested from the earth.

As consciousness is the key to outsmarting waste, please be skeptical when you hear that waste-to-energy facilities are "recycling waste into energy" or anything else that makes burning waste seem like an environmentally beneficial practice. It is simply the propaganda of the industry. By comparison the tobacco industry gained doctors' endorsements up until the 1950s, but you'd be hard pressed to find an MD hawking cigarettes today.[5]

The Upside of Burning

Incineration is hardly the silver bullet that will solve our garbage problem, but all is not lost. There are some inherent pluses in our looking at burning as another way to dispose of the garbage we generate. The rare cases in which an incineration facility produces energy mark a shift in the way people see garbage. Rather than view garbage as a wholly useless output fit for little other than burial, people who harness energy from incineration have at least started to see some value in the waste, albeit a modest one.

Seeing value in garbage is the next step to outsmarting waste. The key questions to ponder are these: How do we gain more value from our waste than just its inherent caloric value? Can all aspects of waste be seen as valuable? And, if so, can we completely eliminate the very idea of waste?

Chapter 5

The Hierarchy of Waste

From the macro perspective, we see garbage as a phenomenal volume of mixed complex materials and primarily useless outputs whose creation is driven by our chronic consumerism. If we are looking for the true value in garbage, it is best to look at garbage from the micro perspective—a perspective that looks at the makeup of garbage—and dissect what a single piece of garbage really is.

Let's use the example of a "disposable" coffee cup. First, it is important to make the distinction between a new and a used coffee cup. Both are coffee cups, but one has positive value: a coffee shop would buy a new cup to serve you that venti chai latte. The other cup—the used one—has negative value (a coffee shop would not buy it back from you). Once you've indulged in your drink, you will most likely put the cup in a garbage can (as coffee cups are not recyclable in today's recycling systems due to the plastic coating on the inside), the owner of which will have to pay for that coffee cup to be transported to a nearby landfill or incinerator.

The Components of a Waste Object

- Composition
- Form
- Intention

The Components of Waste

To see the value in the used coffee cup, we must dissect its physical components and the associated costs of each.

Composition   Every object is made of some physical matter. The used coffee cup, for example, is made primarily of paper, with a thin coating of plastic or wax to ensure that the liquid inside stays inside. It costs money to turn a tree into paper pulp and crude oil into plastic.

Form   The difference between a pile of paper pulp and plastic pellets and a coffee cup is form—a coffee cup is in the shape of a cup. As with the composition, it was a cost to the manufacturer to turn pulp and plastic into the form we know as a cup.

Intention   Finally, the coffee cup is also an idea. There are plenty of different kinds of disposable cups, but not all of them are meant to protect your hands from being scalded by a steaming hot liquid. Just imagine how much research and development time (read: money) it takes to invent anything new.

By looking at where money was spent to develop and produce a new product, one can begin to dissect the value inherent in the used product.

Seeing Value in Waste

If we assume that each cost in the creation of an object is a point of value, we can prioritize the uses of garbage in a simple hierarchy of waste. Of course, there are also things we can do before we make garbage in the first place.

If we really want to outsmart waste, the best place to start is by not creating it to begin with. By consciously reducing levels of consumption, we can eliminate waste in two ways. First, if we don't buy stuff, no waste is created. This option is head and shoulders above any level on a waste hierarchy in terms of sustainability.

Second, buying things that produce useful outputs instead of useless ones also creates zero garbage (unless you put that useful output into a garbage can—remember the 25-year-old hot dogs mummified in landfills?). For example, when you are hungry, consider buying a piece of fruit. Its natural packaging (skin or peel) is a useful output in nature and can easily be composted. By contrast, a bag of candy not only is less healthy but its packaging is a useless output that will harm nature when you throw it away.

Other than avoiding consumption and preventing the creation of useless outputs altogether, the hierarchy of waste is split between two waste solutions: *circular*—reusing, upcycling, and recycling—and *linear*—incinerating and landfilling.

Circular Solutions

Reusing   Even nature leverages reuse on a frequent basis. The small hermit crab can't make its own shells. Snails and mollusks typically manufacture gastropod shells for their own use, but when they die a hermit crab can reuse the shell. As a hermit crab grows, it moves from one scavenged, used shell to another.[1]

Reusing, or buying used, values all three components of that waste object: composition, form, and intention. For example, if you go to a vintage store and buy a used hat, you are valuing the cotton that the hat is made from (composition), the shape and the style of the hat (form), and the fact that the hat was meant to keep your head warm (intention). The best part of reusing, or buying used, is that you support the economy (by buying used stuff) while not voting for a new hat to be manufactured. Although you do not support the economy as much as buying new, buying used is perhaps the most perfect form of consumption in terms of sustainability and environmental impact.

Upcycling   An item is upcycled when one leverages the form and the composition of waste but not the original intention. Upcycling that spent coffee cup would entail using it for something other than serving a hot beverage— perhaps filling it with soil and growing a plant. Sewing together juice pouches to make a backpack or weaving candy wrappers into a bracelet are just a few other examples of leveraging an object's form and composition without using it for its original intention. Upcycling innovation is the greatest among the poorest people in the world, as they don't have the liberty to waste outputs. The more value your mind allows you to see in garbage, the easier it is to unlock the potential of upcycling.

Recycling   When you value only the composition of waste and not its form or intention, recycling comes into play. Recycling a used coffee cup involves shredding it—destroying its original form and any possibility of using it for its original intention in the process. The result of shredding is a mixture of papery pulp and plastic that can be further separated by hydropulping. The resulting paper and plastic—the raw materials it took to make the cup—can then be made into recycled-paper-and-plastic products. The trick to outsmarting waste by recycling it is collecting it in a separated fashion.

Critics of circular solutions say that they are very limited and can't be applied to most forms of waste, but this is simply not true. At TerraCycle we have evaluated every form of consumer waste and found that everything can be either reused, upcycled, recycled, or some combination of those solutions. For example, cigarette butts, used chewing gum, and dirty diapers can be recycled. A chip bag or juice pouch can be upcycled and recycled. Computers, cell phones, shoes, and clothing can be reused, upcycled, and recycled.

reuse   To repurpose a waste object by valuing the material from which it is made, the form that the material is in, and the intention of the form. *Example:* refurbishing a used cell phone so that someone else can use it.

upcycle   To repurpose a waste object by valuing the material from which it is made and the form that the material is in. *Example:* sewing juice pouches together to make a backpack.

recycle   To repurpose a waste object by valuing only the material from which it is made. *Example:* melting soda bottles into carpeting.

And TerraCycle is not the only company using these forms of circular solutions; there are many others around the world that have come up with similar solutions.

Linear Solutions

While reducing consumption doesn't create any waste in the first place, the first three levels of the waste hierarchy rely on circular solutions—solutions that allow constant reuse of the same object (or its parts). The final two levels of the waste hierarchy are linear. Unlike circular solutions, linear solutions can be used only once on any given output. Like the period at the end of this sentence, they represent the definitive end of a product's lifespan.

Incinerating   As it sounds, incineration involves burning trash. Although the vast majority of incinerators produce no energy, a small percentage practice incineration for energy, or waste to energy). Unlike truly circular solutions, incineration for energy leverages only the caloric value of the composition of the material. While incineration for energy is better than landfilling (at least *some* energy comes from it), you can burn something only once; and all of the energy that was used to make the object initially—the energy invested into inventing it, harvesting the raw materials, and then manufacturing a product by giving those materials form—disappears, literally in a cloud of smoke.

Landfilling   Not only can you landfill an object just once but landfilling attributes no value (in fact, negative value) to all of the components of that object. While an object's composition and form can actually be preserved in the

externalities   External effects, often unforeseen, as a result of the production or use of a product. *Example:* air pollution.

largely oxygenless environment of a landfill, they are both pretty useless underground, rendering the object's original intention irrelevant. Outside of harvesting methane emissions for energy (which very few people do), a landfill has only negative externalities—nothing productive comes of mummifying heads of lettuce and leaching toxic chemicals into the earth. A landfill is the embodiment of seeing no value in garbage and treating it solely as useless output.

The Hierarchy of Waste

Circular solutions:

- Reusing
- Upcycling
- Recycling

Linear solutions:

- Incinerating
- Landfilling

The Rules of the Waste Hierarchy

There are two rules to the hierarchy of waste. First, the more you value waste (or the less you do to waste to cycle it again), the better it is for the environment. Reusing is better than upcycling because all of the original components are maintained; upcycling is better than recycling because it maintains an object's form and composition; and recycling is better than linear solutions because the composition of the original product doesn't become a useless output.

Second, an object can go down in the hierarchy only as that particular object. For example, a plastic water bottle can be reused as many times as you wish, but once you have upcycled that bottle into a bird feeder you cannot reuse it to store a liquid. The upcycling process, while maintaining the object's composition and much of its form, destroys the object's original intention by altering its integrity. In the case of the bottle–turned–bird feeder, punching holes in the bottle to mount it as a bird feeder makes it incapable of holding a liquid again (plus, you might not want to drink from a bottle that had bird food in it). On the other hand, once it has been upcycled into a bird feeder, it can be reused *as a bird feeder* indefinitely.

The same principle holds true if you take the bottle (or the upcycled bird feeder) and melt it into plastic by recycling it. Once it is melted, you can't reuse the bottle or upcycle it—both its original intention (required for reuse) and form (required for upcycling) are destroyed. Once you take that melted plastic and make it into a fork, however, it can be reused as a fork many times and that fork can also be upcycled and recycled.

The Rules of the Waste Hierarchy

- The more you value waste, the better it is for the environment.

- An object can go down in the hierarchy only as that particular object.

The Life-Cycle Analysis of the Waste Hierarchy

Today we can measure the environmental impact of a process by conducting a life-cycle analysis (LCA), during which a whole range of environmental impacts—from the amount of carbon produced to the overall impact on an ecosystem—are scientifically measured. LCAs have shown us that the main impact of any product is the result of the extraction of its raw materials from the earth. For example, 90 percent of the environmental impact of a cotton shirt comes from turning soil into a woven cotton textile, while the remaining 10 percent comes from sewing and transporting the shirt. In the case of a bag of chips, it is turning soil into a crispy potato chip and taking oil out of the ground to make the packaging that are mostly to blame.

The hierarchy of waste represents an order of waste solutions from least to most impact on the environment. Reuse is by far the preferred solution because you keep an item from becoming a potential useless output and are not producing any demand for new products to be made. In turn, falling demand for a product eliminates the large environmental cost of extracting materials from the earth and converting them into a new product. Upcycling is next because it typically requires little energy input and can eliminate the need for a new product. Recycling a product by shredding it, separating its components, and making the resulting material into a new product eliminates the need for new materials but requires more energy than upcycling. Both upcycling and recycling typically use far less energy than what it takes to harvest new raw materials from the planet.

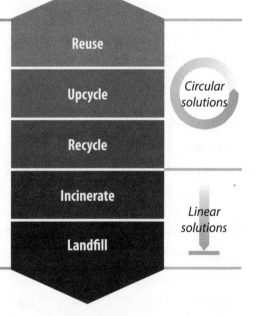

Most sustainable

Reuse

Upcycle

Circular solutions

Managing the garbage we've created

Recycle

Incinerate

Linear solutions

Landfill

Least sustainable

Unfortunately, linear waste solutions like incineration and landfilling don't offset the need for anything new (except perhaps a little energy) and as such are seen as environmental negatives. From an environmental perspective, the more energy we can save when creating goods, the better. And the way to save energy is to stop extracting new materials from the earth and to instead use what we have already extracted.

It is entirely possible to outsmart waste if we look at waste not as a liability but as the confluence of three valuable components: composition, form, and intention. By dissecting the components of garbage, you can see that garbage really doesn't exist—it is just a highly misunderstood resource whose unique characteristics require a new perspective and a little bit of creativity.

Chapter 6

The
Art of
Upcycling

The key difference between upcycling and reusing waste is that with upcycling the original intention of the object changes. For example, if a painter uses a painted canvas for a new painting, he is reusing the canvas. But if instead that same painter takes the canvas apart, uses the wood to make a frame, and uses the fabric to make a purse—that's upcycling.

Upcycling Is Not a New Idea

The idea of upcycling isn't all that new. People have been upcycling for thousands of years. In fact, before the Industrial Revolution (and before processes typically needed for recycling became readily available), reuse and upcycling were common practices. There were no landfills or incinerators to speak of, and the idea of "disposable goods" simply didn't exist in the way it does today. If your pants wore out, the remaining material could be used as cleaning rags or to make another piece of clothing. If a leg broke off your kitchen table, the wood that originally made the table could be used to make a shelf.

The concept of waste is a luxury, and this is perhaps why upcycling is more commonplace in poor countries than in rich ones. If you don't have the resources to buy new objects, you will fulfill your needs by looking at what is available and using that—getting quite creative in the process.

Leave Your Assumptions at the Door

The best way to wrap your head around upcycling is to stop looking at objects as waste. Take a tip from nature and look at your "waste" as a valuable material—an output whose initial intention doesn't need to determine its current purpose. Look at what that object *is* but try to ignore what it *was*. In fact, try to pretend that you don't even know what it was made for in the first place.

For example, from the point of view of upcycling, a chip bag is not food packaging; it's a flexible plastic film. It is a waterproof, colorful, thin, and easy-to-tear material with very high tensile strength. The more obvious applications for such a material are weaving and sewing, but the possibilities are endless.

Take another example: a bicycle chain. If you didn't know it was made as a key component of a bicycle, you would be freed to see it without that lens—as a heavy-duty metal chain that connects to itself and can easily be made into smaller sections. Jewelry, pots, clocks, and a host of other upcycled objects only begin to scratch the surface of the once-a-bicycle-chain's uses.

Or how about a vinyl record? If you didn't know that these objects were made to play music (which may be the case with many younger people), you'd just see a black plastic disc about the size of a Frisbee. If you did some experimenting, you'd find that it can be molded after applying a little heat with a hair dryer. What once might have showered bedrooms and dance clubs with music can now be easily formed into a bowl, a plate, or a clock. This list goes on and on and is really limited only by our imagination.

The Business of Upcycling

People have been upcycling for as long as new objects have broken or, recently, gone out of fashion, but the field is really just picking up from a commercial perspective. In the past decade, socially conscious organizations have made upcycling their business. This first began with various non-governmental organizations (NGOs) in poorer countries like Mexico and the Philippines.

Beginning in September 2004, the People's Recovery, Empowerment, Development Assistance (PREDA) Foundation, a charitable organization that was founded in the Philippines in 1974, began producing, selling, and shipping items made from upcycled juice pouches—a waste stream just as common in the Philippines as in the United States.[1] PREDA trains people to collect used juice pouches (including many students, helping their schools earn money in the process), pays them for their efforts, and teaches them about the environment. After the collected pouches

waste stream   A category of waste, such as plastic cutlery or coffee cups.

are cleaned and sanitized, PREDA makes them available to women who produce handcrafted items. Since 2004 PREDA has sold thousands of upcycled products worldwide.

Mitz, named for a Nahuati word meaning "for you," was founded in 2003 (a year before PREDA) by Judith Achar. She started Mitz to fund Casa de Niños de Palo Solo, a Montessori school in Mexico that has provided community service for low-income students since 1979.[2] Searching for ways to develop the community and have its members share in the work, Judith brought together a group of mothers to manufacture everyday articles from materials thrown out by local schools and business. The primary products are handwoven bags made from various food-packaging wrappers.

Upcycling as an industry is not confined to places like Mexico and the Philippines; people in more-developed regions of the world are also getting into the business. In 1993 graphic designer brothers Markus and Daniel Freitag were on the lookout for a messenger bag. Inspired by the colored truck tarps that they saw on the highways in Switzerland, they started FREITAG, a highly successful company that now makes various upcycled bags available all around the world.

But the list doesn't stop with truck tarps and messenger bags. Chaba Décor upcycles wood from demolished boats and buildings to make picture frames and other items for home decor. Ecoist, similar in function to Mitz, weaves candy wrappers into fashionable bags. Global Exchange upcycles things like old magazines into bowls, flip-flops into doormats, soda cans into wallets, and much more. The Greenshop upcycles print blankets, billboards, and inner tubes to make things like pet collars, laptop covers, and notebook folders. New York–based in2green creates cotton apparel, blankets, and totes using yarn made from T-shirt clippings. Transglass makes beautiful vases and other glass pieces from old wine bottles. Whit McLeod repurposes oak wine casks and barrels to make unique furniture. Trash Amps upcycles soda cans and Chinese-takeout boxes into portable speakers for MP3 players and guitar amps. Upcycle Products repurposes large food barrels into rain barrels and composters. And my company, TerraCycle, a global leader in upcycling, makes everything from cookie-wrapper kites to wine barrel composters.

The list of upcycled products is at least as long as the list of waste product materials that can be used in their construction (and the companies doing it). Fortunately, you don't have to be a company or an NGO to leverage the benefits of upcycling yourself.

Leveraging Upcycling as an Individual

It takes something of a do-it-yourself (DIY) mentality to successfully upcycle at home. Add that to your newfound perspective on waste and, voilà, you are ready to outsmart waste at home by upcycling. Upcycling at home is even more environmentally friendly than having an upcycling company do it because you avoid the environmental impacts that come with transporting waste to the upcycling company and then transporting the upcycled product back to you.

The key to successful at-home upcycling is to first separate out your garbage. Try to keep the organics and the inorganics separate. Even if you don't compost, consider having one garbage can for organic waste and one for inorganics. This simple act of separating will also make it easier in the event that you do eventually start composting.

Once you've started to separate your garbage, consider cleaning it out before putting it into your garbage can, only now your "garbage can" isn't really a garbage can anymore—it's a "raw material storage unit." With this in mind, try to organize it as you would raw materials in a workshop. After you've removed the yogurt from the yogurt cups and the chocolate from the candy wrappers, try to organize the waste into three basic categories.

Flexible packaging   Everything from chip bags to candy wrappers to the notorious plastic shopping bag—if you can crumple it, it fits in the flexible-packaging category. If you keep the material types together (chip bags with chip bags and plastic bags with plastic bags), you can actually fuse them together by putting them between pieces of waxed paper and running a warm iron over them. The resulting material can then be sewn, without tearing, into totes, wallets, lunch boxes, and just about anything you can imagine.

Rigid packaging   Yogurt tubs and plastic bottles are great examples of rigid packaging. Although rigid objects cannot be sewn, they are great building blocks both literally and figuratively. If you punch a drainage hole in the bottom, yogurt cups and margarine tubs can be used to start seedlings or grow full-sized plants. Plastic bottles can be cut in half and made into candleholders or gardening tools and can also make great building blocks for things like sheds, fences, and even homes if filled with sand or something similar.

Everything else   Because flexible and rigid packaging constitutes most home waste, whatever doesn't fit in either category can go together. Because the range of waste types you can encounter here is enormous, "everything else" can be a little more challenging. But don't be alarmed—almost everything has upcycling potential. Wine corks can be made into corkboards, bottle caps into art and jewelry, pens into chandeliers, and on and on. If you are having any trouble thinking of what to create with a particular waste item, go to your favorite search engine, type in the name of the waste stream followed by the word *upcycled,* and you'll likely find a whole slew of ideas.

Categories for Home Waste Upcycling

- Organics
- Inorganic flexible packaging
- Inorganic rigid packaging
- Everything else

With proper separation, a little cleaning, and a DIY spirit, you can effectively upcycle most of your garbage into useful items—and are well on your way to eliminating the idea of waste in your home.

Leveraging Upcycling at Work

Upcycling isn't limited to your home and is a sustainable practice that can also be brought to the workplace. TerraCycle, for example, operates offices in 24 countries, and all of our offices are designed and furnished with entirely upcycled materials. The benefit of outfitting workspaces with upcycled material (beyond having our US headquarters being called one of the "coolest offices in the world" by the *New York Times*) is that it costs much less than traditional interior design options. When you use waste material, your raw material costs are very low, many times free, and sometimes even negative.

Beyond using waste to design an office, many companies are in the business of making stuff. Factories, for instance, almost always generate pre-consumer waste (or "factory waste") in the process of production. Pre-consumer waste is generated not just by the factory at the end of the supply chain but by all of the factories along the entire supply chain and typically represents 3 to 10 percent of annual production.

The great thing about pre-consumer waste is that it is typically clean and best of all it's sorted—it's truckloads of clean labels or clean bottles if you are in the shampoo business, and it's truckloads of trimmings and unused fabrics if you are in the clothing business. Whatever it is, it is prime raw material for upcycling. Consider upcycling the waste your workplace produces, or ask a company that upcycles to help you out.

Don't limit your upcycling ambition to just your home. Outsmarting waste has no boundaries and can be applied individually and organizationally. While there are many ways to upcycle, and a huge number of items can be upcycled, there are a few limitations.

pre-consumer waste   Waste generated by manufacturers. *Example:* fabric scrap from a textile factory.

The Limits of Upcycling

Upcycling is more of an art than a science, and, unfortunately, it does have its limits. First, not everyone is a do-it-yourselfer who is willing to separate and clean waste for upcycling purposes. Not only is upcycling limited by the number of people who are willing to do it but the current market size (as demonstrated by the companies that are in the upcycling business) is very small—perhaps because it's a relatively new business concept or perhaps because the market actually is quite small.

Another limit to upcycling is that it's a relatively low-volume solution compared with the total volume of waste out there. Even if every backpack, pencil case, and tote bag in the world were made out of upcycled juice pouches, that would still represent only about 10 percent of all the juice pouches produced, not to mention that an upcycled backpack covered with logos doesn't appeal to everyone.

Finally, not everything can actually be upcycled. Things like used chewing gum, diapers, and cigarettes cannot technically be upcycled in the same way an old barrel or a plastic bag can.

Regardless of the limits, if you believe in the idea of outsmarting waste by upcycling, you can support it as a practice by buying upcycled goods. This is where that almighty vote you cast by buying stuff comes into play.

Almost any product you can imagine has an upcycled counterpart; it just may take a little time to find it.

For the big-volume solution to garbage, especially the garbage that can't be reused or upcycled, we move down the hierarchy of waste—to recycling.

The Limits of Upcycling

- Not everyone is willing to separate and clean waste.

- It is a low-volume solution compared with the total volume of waste.

- Not everything can be upcycled.

It's a flip-flop replay!
This playground is made
with used flip-flops collected
at Old Navy stores and
recycled by TerraCycle during
Summer 2011. Have fun!

OLD NAVY

designed for **5-12** age group
progressive design playgrounds (800) 585-3131

Chapter 7

The
Science
of
Recycling

Becacuse the waste problem represents more than 11 billion tons per year on a global basis,[1] the fundamental solution to all that waste needs to be on an industrial scale. On such a monumental scale, it is hard to leverage the intention and the form of an object. Reuse and recycling would really have their work cut out for them, and it is much easier to focus just on composition. That is where recycling comes in.

Recycling has been in full force since the dawn of the Bronze Age (3300 BCE). Metals have always been difficult commodities to come by, and there is evidence of bronze and other metals being collected in Europe and melted down for perpetual reuse.[2] This behavior was tied entirely to the economics of waste. It was very expensive to harvest new bronze from rock and significantly easier to just melt down a broken bronze object and make something new.

During the Industrial Revolution, the need for metals was enormous, and metal recycling was in full force. Just imagine how much metal it took to build the railroads that crisscross Europe and North America. During World War II, the US government urged citizens to conserve as much as they could in terms of energy, food, materials, and other essentials. Citizens were encouraged to donate metals to the war effort—helping forge everything from bullets to tanks. The culture of recycling, unlike conservation, stayed in effect after the war.

Due to rising energy costs in the 1970s, there were major investments in recycling. It takes about 95 percent less energy to recycle metals, such as aluminum, than it does to produce virgin material.[3]

The story of recycling glass is largely the same as that of metal (it emerged when the idea of glass emerged), but the story of recycling plastics is quite different.

The Quandary of Plastic Recycling

Partially because plastics are nearly synonymous with cheap, disposable products (and partially because the externalities associated with fossil fuels are not included in their price), the idea of recycling them is not as engrained in our culture as it is for metal and glass. Because those materials are significantly more valuable, they are much more likely to end up reused or upcycled or in a recycle bin.

Even though plastics emerged at the turn of the twentieth century, the first North American plastic recycling programs started only when a number of US states instituted bottle deposit return programs in the 1980s. These programs gained popularity in the 1990s, and today approximately 25 percent of rigid #1 (PET) and #2 (HDPE) plastics are recycled (still rather disappointing).[4] All other forms of plastic—#3 (PVC) to #7 (other)—and all flexible films are not recycled in any meaningful way anywhere in the world. That being the case, only about 5 percent of all plastics in the United States are actually recycled.[5]

What do these numbers actually mean? When you look at any plastic product, you may notice a symbol that looks like a recycling logo with a number in the middle. Though somewhat misleading, these symbols do not mean that the product is recyclable (a blunder of the plastics industry). These plastic identification codes (PICs) were introduced by the Society of the Plastics Industry to provide a uniform system for the identification of different plastic polymer types. PICs help recycling companies separate different plastics for reprocessing, and in most countries all plastic products are required to use them.

Although there is an incredibly large range of plastic types and many new ones invented regularly, there are only seven categories. Plastics #1 and #2 are commonly recycled, but none of the remaining types is recycled in any meaningful way anywhere in the world.

Seven Categories of Plastic by PIC

#1 (PET)   Polyethylene terephthalate is what a soda bottle is made from. Because it is high in value and typically clear in color, and lots of stuff is made from it, it is the most recycled plastic in the world.

#2 (HDPE)   High-density polyethylene is used in everything from water pipes to milk jugs. For the same reasons why #1 plastic is recycled, HDPE is the second-most-recycled plastic in the world: it's high-quality stuff, and there is lots of it.

#3 (PVC)   Polyvinyl chloride is used in things like blister packs, plumbing pipes, and vinyl records. It is not used in food packaging because the plasticizers needed to make natively rigid PVC flexible are usually toxic. PVC is nicknamed the "poison plastic," as each stage in its life cycle—from production to use to disposal—has a negative impact on both the environment and human health. When PVC is melted—which it is both when it's made and later when it's formed into an object—it gives off toxic hydrogen chloride gas that turns into hydrochloric acid on contact with the moisture in our lungs. It is nasty stuff, but we still buy it on a daily basis.

#4 (PE-LD)   Because of its flexibility and strength, low-density polyethylene is popular for use in things like frozen-food bags and squeezable bottles like the mustard bottle in your fridge.

#5 (PP)   Polypropylene is popular packaging for yogurt and margarine as well as material used in many disposable cups and cutlery. Although curbside recycling of PP is very limited, it is an area that is growing slowly in developed countries.

#6 (PS)   Polystyrene is used in similar applications as #5
plastic, and you'll also find things like disposable razors and
CD jewel cases made of the stuff.

#7 (Other)   My personal favorite identifier, a #7 symbol
essentially means "We have no idea what this is." You'll see
this on everything from contact lenses to DVDs.[6]

Plastics Are Recyclable,
So Why Is Recycling So Rare?

Even though all forms of plastic (even mystery #7) can
technically be recycled, its rate of recycling is still quite
low—only 8 percent in the United States, which is a mid-
level recycling market.[7] The biggest difficulty when it
comes to plastic recycling, as with upcycling, is sorting
and separating. While it is easy to recycle #1 or #2 or #3 or
whatever category of plastic on its own, it is very difficult
and costly to recycle when they are all mixed together. The
more we mix plastic types, the lower quality the output.
To compound the problem, many products use multiple
forms of plastic. A simple object like a shampoo bottle is #1
plastic for the bottle and #5 for the cap, a cell phone may be
composed of a dozen different forms of plastic, and your car
probably has parts representing almost all known plastics.

Because oil is relatively cheap and plastics are made from
oil (they are a petrochemical), the value of recycled plastics
is very low compared with the high cost of collecting,

separating, and processing it. The fate of recycling seems to be entirely tied to the economics of its composition.

It's All about the Money

The decision to recycle or not to recycle a certain material has little to do with its technical capacity to be recycled. In fact, all objects in the world can technically be recycled. Instead the decision to recycle or not is tied entirely to economics. Glass, aluminum, uncoated paper, and #1 and #2 rigid plastic are the only five commodities that are readily recycled on a global level because the cost of collecting and processing them is less than the value of the discarded material. There is also a large volume of those materials out there, making it relatively easy to collect meaningful quantities; they can be collected in a separated way or separated later at a recycling center.

All other objects—from used cigarette filters to flexible packaging to dirty diapers—can be recycled, but the cost of collection and processing is greater than the value of the resulting material. Money is the only reason why these materials are not recycled. It's that simple.

But there is some good news. As the price of oil increases, it will force everything to eventually be rendered recyclable. For example, if the price of oil doubled, so would the price of virgin plastic. Suddenly, it would make economic sense to recycle plastics, and you'd likely see recycling rates skyrocket around the world. On the other side of the coin, when oil prices hit a 10-year low in 2009, almost half of the recycling centers in the United States were at risk of shutting down because they simply couldn't find a market for their resulting materials.[8] Many of these plants did cease operations and never reopened.

Certain aspects of outsmarting waste are tied to larger-scale decisions than ones we can influence as individuals. For example, if the United States stopped subsidizing the price of oil (today the US oil subsidy is almost $5 billion per year[9]), it would drive up oil prices and fuel more domestic recycling (no pun intended).

While we wait for oil prices to go up, many folks have started touting the virtues of biodegradable plastic as a way to outsmart waste. But is biodegradable plastic really as smart as it sounds?

> **biodegradable**   A material that can be decomposed by living microbes. *Example:* an apple peel.

The Myth of Biodegradability

Many products, such as a plastic coffee cup, are simply not recyclable in most recycling systems. Therefore, if you are an independent coffee shop (or even a national one), you typically have three choices:

- You can do away with the plastic cup and use another form of packaging that is recyclable or highly reusable (such as glass).

- You can team up with a company like TerraCycle with programs to recycle otherwise "non-recyclable" waste.

- You can use biodegradable cups.

Biodegradable plastics are those made from renewable resources such as corn, sugar cane, and other starchy plant materials. These are refined into a petrochemical equivalent such as polylactide, which has similar properties to #1 plastic (PET) except for the ability to handle heat and the fact that it's capable of being decomposed by bacteria or other living organisms.

The challenge that biodegradable plastic purportedly solves is what happens when it becomes garbage. The obvious benefit of biodegradable plastic is that it has the perceived ability to decompose when it becomes waste. As with many "green" practices, however, the devil is in the details.

When you look at any object, it is important to look at both how it is made and how it is disposed of. It took a lot of energy to turn soil into a plant, more energy to later turn that plant into biodegradable plastic, and even more to shape it into a cup or fork. The most efficient use of that embodied energy would be to make the plant-based plastic durable and keep it as plastic for as long as possible. When that plastic is composted back into soil, all of the energy used to create the cup or fork is effectively wasted.

Even more, if you throw that compostable cup into the trash, it still won't decompose thoroughly in a landfill (remember the mummified lettuce?). Oxygen—which is required for composting—is not available in a landfill, so throwing the biodegradable cup into the trash is basically as bad as trashing a plastic cup. Tossing that biodegradable cup into a recycling bin isn't a good idea either; it still isn't recyclable, and adding it to the mix will actually degrade the quality of the resulting plastic.

How about composting the cup at home? Even if you are willing to carry the cup with you all day back to your home and happen to have a composting pile (as do only about 8 percent of Americans[10]), you will still be out of luck. Most consumer compost piles simply do not get hot enough and are not maintained well enough to decompose biodegradable plastic.

That leaves only one option: in a small number of cities, including San Francisco and Montreal, there is municipal green waste collection that can properly compost biodegradable plastic at industrial composting facilities. In reality, this limits the practical use of biodegradable plastic to a handful of cities. But even if it were possible to compost biodegradable plastic as easily as something like an apple core, at our current levels of consumption there isn't enough farmland in the world to support our plastic needs if we moved entirely to plant-based material.

So why is biodegradable plastic so popular? It's *perceived* to be a silver-bullet solution: we buy a product, and when we throw it away we think it will simply disappear like a banana peel. In actuality it is much more complicated than that. Clearly, the optimal solution is to use materials that can be recycled or reused multiple times, allowing complex materials like plastics to remain complex materials instead of burning them, burying them, or turning them back into soil.

Everything Can Be Recycled

In the end and at a price, every form of plastic can be recycled. Of course, recycling plastic would be more attractive if the price of oil went up, but this economic need can also be made up in the short term. Packaging laws and companies' voluntarily funding recycling systems for their waste would all help the cause.

The key challenge is that our waste, once in a garbage can, becomes mixed together into a complex pile of wide-ranging materials, seriously limiting our ability to do much other than burn or bury it. This only makes separation that much more important.

Chapter 8

The Critical Element of Separation

In nature the waste of organisms is typically spread out in small quantities over a wide area. Unlike humans, animals in nature don't head to the same spot every time they have to poop. They don't preserve their dead, place them in caskets, and later bury them in designated areas. And they certainly don't have any garbage, let alone put it all into a big pile.

When outputs are mixed together as they are in a landfill, it is harder for them to become useful inputs. Putting even useful outputs into the garbage will render them useless outputs. While this is partly because they will not naturally decompose in a landfill, it is also because it is very hard to recycle a soda bottle (#1 plastic) when it is squashed together with a banana, a yogurt cup (#5 plastic), and a used rag.

Any one of these outputs could be recycled or composted individually. A soda bottle could be melted down into plastic, as could a yogurt cup. A banana could be composted, and a rag could be shredded and made into paper or new fabric.

If we do choose to buy products, we must rethink how we accumulate their outputs so as to render them all useful. Think back to the avid upcycler: if you want to successfully upcycle at home, you need to separate your waste instead of mixing it together in a trash can. The same goes for waste at large; while it is convenient to just put all of our outputs into a big pile, it renders them useless in the process.

While the process of outsmarting waste may not be as convenient as what we are doing today, making an extra effort will unlock a host of possibilities that are both environmentally and economically beneficial to us as individuals, to us as a species, and to nature at large.

The Art of Sorting: Before Waste Becomes Garbage

Because waste must be sorted to be processed in a circular fashion (reused, upcycled, or recycled) and to maximize its potential, the question becomes one of how to sort material in a way that is both convenient and cost-effective and—most importantly—how to render our waste valuable, useful outputs. This leaves us with two choices: sort waste before it becomes garbage (the moment you put it into your garbage can) or after it becomes garbage.

Although sorting waste might be less convenient than just tossing it in the trash, it is without question the optimal solution. By sorting at home, you can sort perfectly, relying on your senses to determine in which pile a certain form of garbage should go. Because we largely generate waste one piece at a time, all it takes is putting waste in separate piles instead of mixing it all together.

The easiest place to begin your journey of sorting waste is to separate your organics (like kitchen scraps) and your inorganics (everything else). Even further, you can sort your inorganics by what the local municipal recycling system

Categories for Home Waste Recycling

☐ Organics

☐ Recyclables

☐ Everything else

recycles and what it does not, ultimately leaving you with three piles: organics, recyclables, and everything else. The recyclables you should recycle, but what about the organics and the non-recyclables?

What to Do with Your Organics

There are a few things you can do with your separated organic waste. The first option is to start an indoor (if you live in an apartment) or outdoor composting system. It's not as hard as you might think, and you don't need a lot of space to do it.

Indoor composting Indoor composting systems range from something very simple like a tumbling composter that keeps the compost aerated enough to break down easily, to vermicomposting setups that harness the appetites of red worms to eat the food waste you produce. Personally, I would recommend a red worm system; it is more compact, it doesn't smell, and you have the bonus of a few thousand pet worms.

Outdoor composting   The range of composting systems available for outdoor uses is massive, from simple boxes to rotating systems. The key here is to ensure that the composter has good airflow; aerobic decomposition is critical to prevent your organic food waste pile from becoming your own personal landfill. You can do this by regularly stirring the compost pile with a pitchfork or by using a tumbling composter that enables you to aerate the compost with a hand crank or similar mechanism.

Composting systems attract flies and smell bad only if they are anaerobic (without air), which is a symptom of not turning the compost or adding in too many materials that are not well suited to composting (such as too much meat and oil).

If you prefer not to compost at home, you can always look up the composting facility nearest you. Both www .earth911.org and www.1-800-recycling.com are good online resources. You can also check your local phone directory. Industrial composting facilities typically allow you to deliver your organic waste to them and will often take it off your hands at little or no cost. They may even let you fill up your car with rich compost when you leave.

Pig farmers also love organic waste because it is fantastic feed and will typically accept it for free. A trip to a pig farm could even be a fun adventure, especially if you have kids. You can show them not only what happens with their food scraps but also where bacon comes from.

If you don't want to compost at home or drive your organic waste to an industrial composting facility or pig farm, you always have the choice of throwing it into the landfill. But be conscious that by doing so you are turning an easily useful output into something entire useless (mummified hot dogs, anyone?).

What to Do with Your "Non-recyclables"

Assuming you have decided to recycle your recyclables and compost your organic waste, all that remains is a pile of separated, inorganic, non-recyclable waste—perhaps the hardest pile to deal with. But just like your organic waste, you have some options.

First, you could become an avid home-upcycler. By using this otherwise largely useless material as valuable inputs for your various upcycling endeavors, you keep it out of the landfill and eliminate the need for creating new versions of the products you create in the process.

Second, you could contact a company like TerraCycle that collects this form of waste, often providing free shipping and in some cases a small donation for your favorite nonprofit or school for each piece of garbage that you collect. Over the past decade at TerraCycle, we have found that every type of "non-recyclable" waste in the world can in fact be reused, upcycled, or recycled (or some combination of the three). This is the case for everything from the notorious grocery bag and coffee cup to things that you wouldn't even imagine could be recycled: dirty diapers, chewing gum, hair, used cooking oil, cigarette butts, and even used feminine hygiene products. The critical requirement to outsmarting waste on such a level, as at home, is separation. Separated collection allows each type of waste to be aggregated and industrially processed into new materials.

Dirty diapers, for example, once collected, are put through a gamma radiation process that sterilizes the material, killing harmful pathogens like E. coli and salmonella. The radiated diapers are then shredded, separating the cellulosic material and the superabsorbent polymers (SAP—the stuff that absorbs urine) from the plastics—#4 (PE-LD) and #5 (PP). Composting facilities consider cellulosic material and SAP to be very useful inputs because the former is plant-based and the latter helps compost increase its water retention. The PE-LD and PP are melted and sold as plastic to various companies that consider the material a useful input.

"non-recyclables"   Waste that is made from materials that cannot be recycled in consumer recycling systems. *Example:* a used cigarette butt.

Third, the worst-case scenario, you can simply throw out your "non-recyclables." While this may be the most convenient option, it's also the one that does the most harm to the environment and ultimately contributes to our ever-growing garbage problem. If we can outsmart even diaper waste and make it into entirely useful outputs, *every* type of waste can and should be outsmarted. It's all possible—it just takes a little extra effort.

Options for "Non-recyclables"

- Become an avid home-upcycler
- Work with a company like TerraCycle
- Throw out your "non-recyclables" (and be part of the problem)

The Science of Sorting: After Waste Becomes Garbage

Although sorting before your waste hits the trash can is certainly easier, sorting can also happen once waste is mixed together. Practically speaking, this happens almost exclusively with recyclable materials because it's impossible to economically sort mixed garbage. Many recycling systems today allow you to combine your glass, aluminum, paper, and plastics (#1 and #2) for the sake of consumer convenience. In the industry this is called single-stream recycling.

single-stream recycling Consumer recycling system in which glass, aluminum, paper, and plastics are combined for the sake of convenience.

After your recyclables are picked up and taken to a municipal recycling facility, the pile is either manually or automatically sorted. In a manual system, the material is put onto conveyor belts, and workers take out the recyclable materials by hand. Everything else (sometimes up to half of the stuff we put in our recycling bins) ends up being sent from the recycling center to a landfill. If material is automatically sorted, the waste is sent through a range of systems, from magnetic separation (which takes out the metals) to methods that sort based on the density (water separation) or composition (optical separation) of the materials.

Whether manually or automatically separated, a significant percentage of the material you thought was going to be recycled still ends up in a landfill. For example, if you throw out a window cleaner bottle without taking off the trigger head, the entire bottle will be thrown out. Although the bottle is made from #1 (PET) plastic, which is recyclable, the trigger head is made from a variety of plastics and metals. The same goes if you throw out a shampoo bottle: the bottle, like the window cleaner, is #1, but the cap is #5.

It's simply not economical to manually remove the trigger head or cap so that the bottle can be recycled, so it all ends up in the landfill.

No matter what system or combination of systems we rely on, the most important step in outsmarting waste on any level is sorting it. It is significantly easier to circularly process material once it is separated.

Can We Solve the Problem before the Need to Sort?

As always, the simplest way to avoid this whole mess to begin with is to buy differently. If you need to buy, perhaps the most important thing to consider is whether the products end up as useful or useless outputs.

When buying (or, by extension, when making) products, look for ones that use exclusively recyclable materials: glass, aluminum, paper, or #1 and #2 plastics. Avoid items that combine material types (whether recyclable or not) because these will typically not be recycled.

For example, perfume is typically packaged in a fancy glass bottle with a pump that is crimped onto the glass, making it very difficult to remove. Even though the glass, on its own, and the metal, on its own, can be recycled, together they make for a useless output. The same goes for a host of other products. This also applies to recyclable material that is coated. From the paper cup with a thin plastic coating to the plastic toothpaste tube lined with aluminum, if you coat otherwise recyclable material, it becomes non-recyclable in our current recycling systems.

If you take the time to really understand the waste you produce, you will almost undoubtedly be shocked (as I was when I first started) by the amount of stuff we throw out. When you think about sorting, don't see it as a burden. In essence it is the critical cornerstone of outsmarting waste, and it's the first step toward living in a world that doesn't see waste as garbage—a negative—but rather as a positive, as nature does.

Make a Difference

TERRACYCLE®

Recycle These Cups!

see details on back

squared

50

Chapter 9

The Economics of Out-smarting Waste

The main reason why waste is sent to landfills and incinerators and why few of our outputs are recycled (like they technically can be) is all tied up in the economics of waste. It is simply more expensive to collect and recycle most things than the results are worth, and it's cheap—because we allow it to be cheap—to send waste to a landfill or an incinerator.

Because our world is so economically motivated, perhaps we can make outsmarting waste more attractive by speaking the language of economics. There are hidden economic benefits of investing in the process of outsmarting waste on several different levels. Like the whole of outsmarting waste, these benefits can begin with you at home.

The Benefits of Outsmarting Waste on an Individual Level

Although outsmarting waste may require an investment of your time, every aspect should save you money. If you don't buy unnecessary items, you can save money for something more important. Packaged processed food tends to be more expensive then unpackaged fresh foods. Durable products, even though they may cost more initially, will last longer than disposables and should save you money over the long term. Buying used instead of new will also leave a few extra bucks in your pocket.

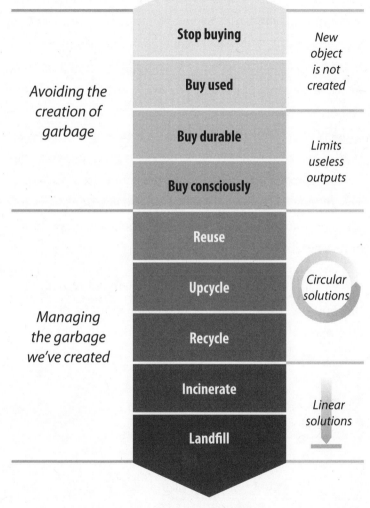

Most sustainable

Avoiding the creation of garbage

Stop buying

Buy used

New object is not created

Buy durable

Buy consciously

Limits useless outputs

Reuse

Upcycle

Circular solutions

Managing the garbage we've created

Recycle

Incinerate

Linear solutions

Landfill

Least sustainable

Outside of buying differently, you can save money if you compost your organics, recycle your inorganics, and upcycle your non-recyclables. This will take more work than putting everything in a garbage can, but you will see a number of economic gains. Your waste management bill, if you are billed by weight, will decrease or completely disappear. You won't have to buy potting soil, as you'll be making fantastic compost. In states with container deposit laws, you will actually get money back for bringing your empty beverage bottles and cans to a recycling center. And because you will be able to make upcycled goods from your waste, you'll save money by buying less stuff overall. If you don't want to upcycle at home, look for services that help you recycle your non-recyclables.

You will also have the noneconomic benefit of knowing that you are not only contributing to solving a major environmental problem but actively working to fight it.

Ways to Save Cash While Outsmarting Waste

- Buy differently
- Compost your organics
- Recycle your select inorganics
- Upcycle your non-recyclables

The Benefits of Outsmarting Waste on a Corporate Level

The economic benefits of outsmarting waste at a broader, institutional level include everything you gain from outsmarting waste at an individual level and then some.

When it comes to offices themselves, you can utilize upcycled and recycled waste in your design and furnishings. Used doors make for great desks, vinyl records make for great wall dividers, and shredded wine corks and flip-flops make for a fantastic walking surface. This is a concrete way to show your employees that the company they work for cares about environmental issues. As a side benefit, you'll have a more funky office environment with a lower overall cost than with traditional interior design options.

If your company manufactures products, you could find recycling and upcycling companies to take your pre-consumer waste materials for use as inputs in their processes. You could even go a step further and use upcycled or recycled materials in the products your company produces. If you manufacture clocks, why not look at making a line out of old CDs and DVDs? If you manufacture chairs, why not try building some from old skis? Consider looking at integrating recycled content into your products. The possibilities are as endless as your imagination, and leveraging raw materials that often have a negative cost can only help a company's bottom line.

But what if you don't need to redesign your office space or
don't want to make your products using waste? Creating
collection programs through which consumers can collect
your non-recyclable product waste and send it back to you
is another way to outsmart waste at an institutional level
while building positive relationships with your customers
and clients. Many companies, from Estee Lauder[1] to Nike,[2]
run voluntary extended product responsibility (EPR)
programs in the United States and around the world.
Offering free shipping can make these programs financially
appealing to consumers and also decreases the likelihood of
your products ultimately ending up in a landfill.

Ways Companies Can Outsmart Waste

- Use upcycled and recycled items in office
  design and furnishings

- Offer pre-consumer waste for upcycling or
  recycling

- Offer an upcycled or recycled product

- Create an EPR program to motivate
  consumers to recycle

- Educate consumers about what is and
  isn't recyclable and provide collection
  facilities

If you aren't in a position to run an EPR program yourself, you could join the hundreds of companies that work with TerraCycle to collect and repurpose waste, saving it from the waste stream and eliminating the need to extract many new resources from the earth each year.

Unique Services That Help Outsmart Waste

Many companies leverage waste to create concrete environmental benefits. Recyclebank, for example, incentivizes people by awarding points if they increase their recycling rates. These points can be redeemed for gift cards, coupons, and various other perks. The platform has yielded benefits—like increased market share—to those companies that sponsor the points (such as Ziploc and P&G) and has increased recycling rates tremendously in communities that have brought in the system, saving millions of dollars for local municipalities.

Besides motivating consumers to participate in recycling programs, companies can take concrete actions to lessen their environmental impact, such as educating the public about what is and isn't recyclable and providing collection facilities. Earth911 has become incredibly successful by providing robust information to consumers about what is recyclable and where it can be recycled. Preserve is a company that makes toothbrushes and plates from #5 plastic collected at various supermarkets (primarily Whole Foods stores).

While there are a number of services out there, the field
is still very limited—perhaps because people don't really
like garbage and generally see the business of waste
management as unsexy.

Extended Product Responsibility Laws

Is there a role for government in outsmarting waste?
Instead of voluntary EPR programs (such as TerraCycle has
innovated), should there be an expansion of mandatory
EPR laws? Should other laws or policies require producers
to take more responsibility for the waste that their products
generate and for the environmental impact associated with
that waste? The concept of extended product responsibility
was first formally introduced by Thomas Lindhqvist two
decades ago. In a 1990 report to the Swedish Ministry of
Environment, Lindhqvist said: "[EPR] is an environmental
protection strategy to reach an environmental objective of
a decreased total environmental impact of a product, by
making the manufacturer of the product responsible for the
entire life cycle of the product and especially for the take-
back, recycling and final disposal."[3]

The system first emerged in Germany, under the Green
Dot, or Grüner Punk. Duales System Deutschland GmbH
originally introduced the system in 1991 following changes
to the Waste Act mandating that companies pay an EPR
fee to Grüner Punk. Similar Green Dot systems were
subsequently adopted in 23 other European countries

and are now used by more than 130,000 companies—
representing about 0.5 trillion packages per year.[4]

The basic idea of Green Dot systems is that manufacturers
pay into a fund for each package they produce. The fund
then administers the money to waste management
companies that actually collect and process the waste
ultimately resulting from these products. When consumers
see the Green Dot logo on a product, they know that the
product's manufacturer has contributed to the cost of
recovery of that product. Many people have come to think
that the logo means the product will be recycled, but
unfortunately that is generally not the case.

Often the money paid by manufacturers goes into a fund
that is administered by a group like the Duales System in
Germany, Eco-Emballages in France, Chepko in Turkey,
and TMIR in Israel. These groups are then charged with
spending that money to recycle as much as possible, and
they almost always focus entirely on increasing the recycling
rate of traditional recyclable materials such as glass and
paper. Although the increase in overall recycling that
typically occurs is a big win for these laws, non-recyclables
are still usually incinerated because recycling non-
recyclables is much more expensive than recycling readily
recyclable materials.

While EPR laws are able to extract billions of dollars from
companies as a tax on their packaging waste, they still don't
do much in terms of recycling traditionally "non-recyclable"

materials. In fact, these laws sometimes have a negative impact on the ability to bring recycling solutions to hard-to-recycle materials because they seem to encourage fallacies like classifying incineration as "recycling into energy" as a way to cheaply comply with the law.[5] In other words, in most EPR programs around the world incineration of waste is classified as legitimate "recycling." This is perhaps one of the most problematic challenges to outsmarting waste because "waste to energy" is a linear solution, not a circular one.

Someone once asked me what I would do if I were given the keys to the national EPR system in a country. Instead of a flat tax on all packaging, I would tax all packages and products on the true cost of their collection and repurposing. For example, a diaper package, as it is a simple plastic and lightweight, would cost only a few cents to collect and recycle. A dirty diaper, however, may cost more than the actual cost of the diaper. With that said, something that put value into the system, such as an aluminum can, would get you a credit. And something like a gold watch would get you even more. If this is how EPR were set up, the true costs of dealing with the waste that our packages and products eventually become would be billed to the manufacturers. And, just perhaps, it would create the necessary fiscal incentives to push manufacturers to make products that are easier to circularly process or, ideally, add value back into the system.

Taxing the Full Cost of a Product

If EPR laws can successfully tax companies for the waste their products produce, it should be possible to fully tax companies for all of the externalities of their existence— from the materials they extract from the earth to the fumes they pour into the atmosphere. Taxing the full cost of a product would essentially force sustainability onto all corporations, whether or not they consider themselves "green." There is already a model for this program in the way that we tax cigarettes. Given the known dangers associated with smoking, anywhere from $0.17 per pack in Missouri to $4.35 per pack in New York[6] is collected to help offset the costs incurred by hospitals treating people who develop lung cancer.

If we started taxing the whole cost of a product the way we tax cigarettes, the positive environmental impact would be almost immediate. Because the cost of raw materials would be much higher, companies would have to start making products in a way that would allow them to cycle more of the material they convert into their product. For example, we would pay the true (fully loaded) cost of oil instead of the subsidized cost of just extracting it from the environment and refining it. As the cost of raw materials rose, so would the desire to cycle used material. If this happened, companies would have to start looking at waste very differently.

Increasing the cost of landfilling or incineration with taxes or by charging higher fees to access these disposal methods would have a similar effect. The higher the cost of disposal goes, the more people would look at material recovery as a way to gain the needed inputs for their products. Garbage as a concept would almost cease to exist in industry because it would cost more to dispose of waste in a landfill than it would to reuse, upcycle, or recycle it.

Outsmarting waste can be forced onto companies and the public by increasing the cost of linear disposal (landfilling and incineration) or of extracting virgin raw materials (extended product responsibility). Both can be accomplished by taxes and regulation—or, inevitably, by time. As landfills fill up, the cost of opening more of them will increase, and as oil becomes scarcer, the cost of extracting it in harder-to-reach places (and in smaller quantities) will drive up its cost, as well.

In several industries today, I am seeing manufacturers looking ahead and investing to obtain access to the materials stream generated by their waste; they realize that the economics are evolving to the point where producing new materials will at some point be more expensive than collecting and recycling used materials. It is also an interesting way to gain control over the raw materials on which your product depends.

Environmental laws and taxes speed up the inevitable cost increase and allow us to conserve our valuable space and fossil fuels. At the same time, these laws and taxes force us to discover newer, cleaner, and more-efficient ways of producing and consuming. Given the finite amount of space and resources on our planet, these cost increases *will happen*, over time, whether we like it or not. The question is whether we will choose to slow the process of resource depletion with regulation or continue to use up what few resources we have at a blinding pace, driven by growth-focused economics. This is where the vote you cast at the ballot box really comes into play in the effort to outsmart waste.

If Big Tobacco Can Outsmart Waste, Any Industry Can

When companies have invested in voluntary EPR systems, the results have been tremendous. The most extreme example of an EPR system is perhaps TerraCycle's work with the cigarette industry.

When TerraCycle launched the world's first national cigarette recycling program in Canada (and months later in the United States and Spain), the public reacted incredibly positively. Within just months of launching in Canada, the TerraCycle platform was collecting and recycling more than 1 million used butts per month. Around the same time, the

thousandth positive article praising tobacco companies
for taking responsibility over their waste was printed. Since
then we have set up similar programs in many countries
around the world, from Argentina to Hungary.

Cigarettes are probably the most hated consumer product
in mass circulation. Ever since we discovered that cigarettes
cause cancer, there has been almost no positive publicity
for the tobacco industry. In fact, movies like *Thank You
for Smoking* highlight the challenges that the lobbying
industry has when dealing with tobacco. Jokingly, the
protagonist in the movie, a tobacco lobbyist, calls himself
a "merchant of death."

On top of the known negative health impacts, cigarettes
are the most littered item in the United States; about 135
million pounds of butts are tossed on roadways, thrown in
the trash, or put in public ashtrays every year.[7] According to
a 2009 study, cigarette waste accounted for 38 percent of
all US roadway litter.[8] As part of the Ocean Conservancy's
annual one-day International Coastal Cleanup, more than
2 million cigarettes or cigarette butts—enough to fill
more than 100,000 packs—were removed from American
beaches and inland waterways in 2011.[9] If an industry
whose products literally kill people can benefit from
outsmarting waste, just think of the impact it can have for
far less lethal industries.

Perhaps Nicolas Mallos, a marine debris specialist, said it best when he quipped: "Trash is really too valuable to toss, so we need to find alternative ways to repurpose it."[10] If the language of economics is the one our world really understands, perhaps it is time we start to speak it—with our ballots and our wallets—until the concept of garbage is a thing of the past.

Conclusion

Waste Is Over! (If You Want It)

Disposable goods are cheap to purchase, even cheaper to manufacture, and relatively inexpensive to burn or bury when we're finished with them. If we lived on a planet with infinite space and raw materials, and if economic metrics were the only ones that really mattered in judging the health and the happiness of a species and our planet, a disposable consumer culture just may work. We could extract infinite fossil fuels to make infinite, disposable plastic products to eventually bury in our infinitely expanding landfills, and no harm would be done. The reality, however, is that we live in a place with both finite space and finite resources, and the global garbage crisis grows with our collective appetite.

It hasn't always been this way, and it doesn't have to stay this way. The good news is that there isn't a product we create that can't somehow be reused, upcycled, or recycled, and the idea of trash—of a useless output—is one with no basis in nature. We got ourselves into this mess (no pun intended), and we're the only ones who can get ourselves out of it.

So, what now? There are more than a few things that can be done to outsmart waste.

Buy Differently

Outsmarting waste is a journey that begins with us as individuals, and it starts with the choices we make as consumers. Let's change how we buy things by first questioning the very act of buying.

Before you make your next purchase, ask yourself if you really need the item in question. By purchasing only items we really need, we minimize the amount of waste we create (and save money in the process). When not buying something isn't an option, we can have the smallest impact on the earth by buying used durables—keeping a perfectly functional product from the waste stream after its initial owner discards it while at the same time contributing our dollars to the secondhand economy. If something must be new, always try to buy durable rather than disposable goods.

Be conscious of the outputs that are created with each purchase you make. Bringing your own reusable bag to the grocery store and buying fresh foods that don't have packaging can keep a pile of plastic out of a landfill over time (and buying fresh is better for your health too). Buying one ceramic coffee mug can keep hundreds of paper ones out of the garbage. Just a small change to your consumer habits can have a measurable impact in no time.

See Waste as Value

Hopefully by now, your perspective on garbage has changed somewhat, and perhaps you have begun to see positive value where there was only negative before. There is no concept of trash in nature, and putting your waste to good use only helps bring us closer to that natural equilibrium where one organism's output is another's useful input.

When it comes to the hierarchy of waste disposal, reuse is always the best option—it leverages all of the energy that was needed to make the product and makes use of its form, composition, and original intention. Even if you no longer have a use for a particular item, if it still serves its original function, give it to someone who can use it. If that is not an option, sell it at a consignment store or donate it to a local thrift shop.

Reuse isn't always a viable option, but there is still a lot that you can do to minimize the amount of useless outputs you create. Compost at home, participate in your local recycling program, and consider doing some upcycling—you can even expand these practices by taking them to your school or workplace. Acknowledging the value inherent in our waste makes it that much easier to keep it out of the trash can.

Vote Early and Vote Often

Beyond the impact each purchase can and does have on the environment, it represents an opportunity to voice our concerns and values in the market. Each time you buy something, you are voting for more of that thing to be created. When you buy a pack of disposable plates, it sends the message that more should be made and that disposable plates are something that should exist. The checkout line is at least as powerful as the voting booth, and every time you make a purchase, you are punching an economic ballot. Wield your dollars wisely.

When it comes to voting in actual elections, we have another chance to put in place people working to make garbage a thing of the past. Lobbying local representatives, working to instate municipal recycling and composting programs, and supporting candidates with similar values are just a few of the ways you can work to outsmart waste on a larger scale.

Waste Is Over

If we view the world through the lens of nature, we find that outsmarting waste is not only in our economic best interest but also in the best interest of our planet and fellow co-inhabitants. Completely outsmarting waste is entirely up to our imagination and appetites. It begins on the individual level but is something we need to take seriously on a national and global scale. We just need to see the value inherent in what we discard.

Accepting or deciding to outsmart waste is a choice. The first step is to understand that waste isn't inevitable and to accept the elimination of the idea of waste as a ripe objective for our time. It helps to see the waste equation through the lens of nature, as linear solutions will become objectionable and the migration of our production and consumption to circular solutions will come into focus.

Waste is over (if you want it), and even one change in your buying or disposal habits is a step in the right direction.

Notes

Introduction

The Unique Nature of Garbage

1. US Environmental Protection Agency, "Municipal Solid Waste," http://www.epa.gov/epawaste/nonhaz/municipal/index.htm (accessed July 11, 2013).

2. Greenpeace, "Plastic Debris in the World's Oceans," http://www.unep.org/regionalseas/marinelitter/publications/docs/plastic_ocean_report.pdf (accessed July 11, 2013).

Chapter 1

Where the Modern Idea of Garbage Originated

1. Reuseit, "Facts about the Plastic Bag Pandemic," http://www.reuseit.com/learn-more/top-facts/plastic-bag-facts (accessed July 11, 2013).

2. Connecticut Plastics, "The Men Who Invented Plastic," http://www.connecticutplastics.com/resources/connecticut-plastics-learning-center/the-men-who-invented-plastic (accessed July 11, 2013).

3. United Nations Environment Programme, "Action Urged to Avoid Deep Trouble in the Deep Seas," http://www.unep.org/Documents.Multilingual/Default.asp?DocumentID=480&ArticleID=5300&l=en (accessed July 11, 2013).

4. Inter-Organization Programme for the Sound Management of Chemicals, "State of the Science of Endocrine Disrupting Chemicals—2012," http://www.who.int/ceh/publications/endocrine/en (accessed July 11, 2013).

5. Reference for Business: Encyclopedia of Business, 2nd ed., "Earl Tupper: Farmer and Tree Surgeon, An Interest in Plastics," http://www.referenceforbusiness.com/businesses/M-Z/Tupper-Earl.html#b (accessed July 11, 2013); Chemical Heritage Foundation, "Nylon: A Revolution in Textiles," http://www.chemheritage.org/discover/media/magazine/articles/26-3-nylon-a-revolution-in-textiles.aspx (accessed July 11, 2013).

6. Product Policy Institute, "Unintended Consequences: Municipal Solid Waste Management and the Throwaway Society," http://www.productpolicy.org/ppi/attachments/PPI_Unintended_Consequences.pdf (accessed July 11, 2013).

7. Institution of Mechanical Engineers, "Global Food: Waste Not, Want Not," http://www.imeche.org/docs/default-source/reports/Global_Food_Report.pdf?sfvrsn=0 (accessed July 7, 2013).

8. Ibid.

9. Ask Nature, "Hydrophobic Surface Allows Self-Cleaning: Sacred Lotus," http://www.asknature.org/strategy/714e970954253ace485abf1cee376ad8 (accessed July 11, 2013).

10. Biomimicry Institute, "Learning Efficiency from Kingfishers," http://www.biomimicryinstitute.org/case-studies/case-studies/transportation.html (accessed July 11, 2013).

Chapter 2

The Role of the Individual Purchase

1. Bernard London, *Ending the Depression through Planned Obsolescence* (Madison: University of Wisconsin, 1932).

2. Wisconsin Historical Society, "Brooks Stevens," http://www.wisconsinhistory.org/topics/stevens (accessed July 11, 2013).

3. Statistic Brain, "Light Bulb Statistics," http://www.statisticbrain.com/light-bulb-statistics (accessed July 11, 2013).

4. Good Magazine Online, "The Conspiracy to Control the World's Electric Lighting," http://www.good.is/posts/the-conspiracy-to-control-the-world-s-electric-lighting (accessed July 11, 2013).

5. Greenwashing Lamps, "The Lightbulb Conspiracies," http://greenwashinglamps.wordpress.com/tag/fluorescent-tube (accessed July 11, 2013).

6. Bankrupt.com, "Headlines" *Class Action Reporter* 6, no. 161 (2004), http://bankrupt.com/CAR_Public/040816.mbx (accessed July 11, 2013).

7. Giorgos Kallis, Christian Kerschner, and Joan Martinez-Alier, "The Economics of Degrowth," *Ecological Economics* 84 (2012): 172–80; http://www.samfak.gu.se/infoglueCalendar/digitalAssets/1781245508_BifogadFil_Kallis_2012_The-economics-of-degrowth.pdf (accessed July 11, 2013).

8. Gandhian Institute Bombay Sarvodaya Mandal and Gandhi Research Foundation, "Gandhi Quotes: Hunger," http://mkgandhi.org/epigrams/h.htm#Hunger (accessed July 11, 2013).

Chapter 3

Our Primary Global Solution to Waste: Bury It

1. Russell Hardin, "Garbage Out, Garbage In," *Social Research* 65, no. 1 (1998): 9-30; http://www.nyu.edu/gsas/dept/politics/faculty/hardin/research/GarbageOut.pdf (accessed July 11, 2013).

2. Greenpeace, "Plastic Debris in the World's Oceans," http://www.unep.org/regionalseas/marinelitter/publications/docs/plastic_ocean_report.pdf (accessed July 11, 2013).

3. Ibid.

4. US Environmental Protection Agency, "Decision Maker's Guide to Solid Waste Management, vol. II," Chapter 7: Composting, http://www.epa.gov/osw/nonhaz/municipal/dmg2/chapter7.pdf (accessed July 11, 2013).

5. Ibid.

6. William Grimes, "Seeking the Truth in Refuse," *New York Times,* August 13, 1992, http://www.nytimes.com/1992/08/13/nyregion/ seeking-the-truth-in-refuse.html?pagewanted=all&src=pm (accessed July 11, 2013).

7. Ibid.

8. People Investigating Toxic Sites, "Explosions and Fires at Dumps (Landfills)," http://www.toxicsites.org/Explosions_And_Fires_At_ Dumps.pdf (accessed July 11, 2013).

9. Ibid.

10. C. Bartone, "The Value in Wastes," *Decade Watch* (January 1988): 3–4.

11. Bharati Chaturvedi, "Mainstreaming Waste Pickers and the Informal Recycling Sector in the Municipal Solid Waste" [research paper], *Handling and Management Rules 2000, A Discussion Paper* (2010).

12. International Labour Office/International Programme on the Elimination of Child Labour, "Addressing the Exploitation of Children in Scavenging (Waste Picking): A Thematic Evaluation of Action on Child Labour (2004), www.ilo.org/ipecinfo/product/ download.do?type=document&id=459 (accessed July 11, 2013).

13. Mayling Simpson-Hebert, Aleksandra Mitrovic, and Gradamir Zajic, *A Paper Life: Belgrade's Roma in the Underworld of Waste Scavenging and Recycling* (Loughborough University: Water, Engineering and Development Centre, 2005).

14. David Geeter, Matthew Mitsui, Benjamin Lu, Conrad Peterson, Alexander Blumenthal, and Adam Schatteman, "What a Waste: Designing a Solution for New Jersey's Waste Management Crisis," Rutgers University, http://soe.rutgers.edu/sites/default/files/gset/ Solid.pdf (accessed July 11, 2013).

15. Global Alliance for Incinerator Alternatives, "Incinerators: Myths vs. Facts about 'Waste to Energy,'" http://www.no-burn.org/downloads/ Incinerator_Myths_vs_Facts%20Feb2012.pdf (accessed July 14, 2013).

16. Keep Oklahoma Beautiful, "21st Century Landfills," http://www
.keepoklahomabeautiful.com/21st-century-landfills (accessed
July 11, 2013).

Chapter 4
The Energy Inherent in Our Waste

1. Window on State Government: Susan Combs, Texas Comptroller of
Public Accounts, "Chapter 18: Municipal Waste Combustion," http://
www.window.state.tx.us/specialrpt/energy/renewable/municipal
.php (accessed July 11, 2013).

2. Global Alliance for Incinerator Alternatives, "Incinerators: Myths vs.
Facts about 'Waste to Energy,'" http://www.no-burn.org/downloads/
Incinerator_Myths_vs_Facts%20Feb2012.pdf (accessed July 14,
2013).

3. California Environmental Protection Agency Air Resources Board,
"Glossary of Air Pollution Terms," http://www.arb.ca.gov/html/gloss
.htm (accessed July 11, 2013).

4. "Wisconsin State Representative Louis J. Molepske Jr., 71st
Assembly District, Testimony on Assembly Bill 372 for the Senate
Committee on the Environment, March 16, 2010," http://legis
.wisconsin.gov/lc/comtmats/old/09files/ab0372_20100318131840
.pdf (accessed July 11, 2013).

5. Martha N. Gardner and Allan M. Brandt, "The Doctors' Choice Is
America's Choice: The Physician in US Cigarette Advertisements,
1930–1953," *American Journal of Public Heath* 96, no. 2 (2006);
http://www.ncbi.nlm.nih.gov/pmc/articles/PMC1470496 (accessed
July 11, 2013).

Chapter 5
The Hierarchy of Waste

1. BBC Nature Wildlife, "Hermit Crabs," http://www.bbc.co.uk/nature/
life/Hermit_crab (accessed July 11, 2013).

Chapter 6

The Art of Upcycling

1. Academy for Educational Development, "Philippines NGO Recycles Aluminum Drink Pouches to Raise Money," http://www.human trafficking.org/updates/129 (accessed July 11, 2013).

2. Mitz, "History," http://mitz.org.mx/nosotros (accessed July 11, 2013).

Chapter 7

The Science of Recycling

1. United Nations Environment Programme, "Waste: Investing in Energy and Resource Efficiency," http://www.unep.org/greeneconomy/Portals/88/documents/ger/GER_8_Waste.pdf (accessed July 11, 2013).

2. Pre-Industrial Recycling, "Case Study: Metalworking and Recycling in Late Bronze Age Cyprus," http://www.preindustrialrecycling.com/bronzeage.html (accessed July 11, 2013).

3. Earth911, "Facts about Aluminum Recycling," http://earth911.com/recycling/metal-3/aluminum-can/facts-about-aluminum-recycling (accessed July 11, 2013).

4. Eartheasy, "Plastics by the Numbers," http://eartheasy.com/blog/2012/05/plastics-by-the-numbers (accessed July 11, 2013).

5. Green (Living) Review, "American Plastics Council Estimates Only about 5% of All Plastics Are Being Recycled," http://greenreview.blogspot.com/2011/02/american-plastics-council-estimates.html (accessed July 11, 2013).

6. Earth911, "Recycling Mysteries: Plastic #7," http://earth911.com/news/2009/11/30/recycling-mystery-plastic-7 (accessed July 11, 2013).

7. US Environmental Protection Agency, "Plastics," http://www.epa.gov/osw/conserve/materials/plastics.htm (accessed July 11, 2013).

8. Kate Galbraith, "Recycling Market Plummets," *New York Times,* November 7, 2008, http://green.blogs.nytimes.com/2008/11/27/recycling-market-plummets (accessed July 11, 2013).

9. Robert Rapier, "The Surprising Reason That Oil Subsidies Persist: Even Liberals Love Them," *Forbes,* April 25, 2012, www.forbes.com/sites/energysource/2012/04/25/the-surprising-reason-that-oil-subsidies-persist-even-liberals-love-them (accessed July 11, 2013).

10. Earth911, "Compost Awareness Week: No More Excuses, Start Your Pile," http://earth911.com/news/2010/05/03/compost-awareness-week-no-more-excuses-start-your-pile (accessed July 11, 2013).

Chapter 9
The Economics of Outsmarting Waste

1. Recycle Caps with Aveda: http://www.aveda.com/cms/discover_aveda/bethechange/popup_caps.tmpl.

2. Nike Reuse-a-Shoe: http://www.nike.com/us/en_us/c/better-world/stories/2013/05/reuse-a-shoe.

3. Institute for Local Self-Reliance, "The Concepts of Extended Producer Responsibility and Product Stewardship," http://www.ilsr.org/the-concepts-of-extended-producer-responsibility-and-product-stewardship (accessed July 11, 2013).

4. Jessica Salter, "Green Arrow Symbol Does Not Always Point to Recycling," *London Daily Telegraph,* August 31, 2008, http://www.telegraph.co.uk/news/uknews/2656458/Green-arrow-symbol-does-not-always-point-to-recycling.html (accessed July 11, 2013).

5. Conrad MacKerron, "Unfinished Business: The Case for Extended Producer Responsibility for Post-Consumer Packaging," http://asyousow.org/publications/2012/UnfinishedBusiness_TheCaseforEPR_20120710.pdf (accessed July 11, 2013).

6. Centers for Disease Control and Prevention, "State Cigarette Excise Taxes: United States, 2010–2011," http://t.cdc.gov/synd.aspx?js=0&rid=cs_1051&url=http://t.cdc.gov/MA2 (accessed July 11, 2013).

7. Clean Virginia Waterways, "How Many Discarded Cigarette Butts Are There?" http://www.longwood.edu/cleanva/cigbutthowmany.htm (accessed July 11, 2013).

8. Keep America Beautiful, "Litter Prevention," http://www.kab.org/site/PageServer?pagename=focus_litter_prevention (accessed July 11, 2013).

9. Ocean Conservancy, "International Coastal Cleanup 2012 Data Release," http://www.oceanconservancy.org/our-work/marine-debris/check-out-our-latest-trash.html (accessed July 11, 2013).

10. Michael Felberbaum, "Reynolds Subsidiary Funding Cigarette Recycling," Associated Press, http://bigstory.ap.org/article/reynolds-subsidiary-funding-cigarette-recycling (accessed July 11, 2013).

Index

About Tom Szaky

© TerraCycle

Tom Szaky was born in Budapest in the early 1980s during a time of communism and economic hardship. After the nuclear power reactor in Chernobyl, Ukraine, exploded in 1986, Tom's family was able to escape Hungary and move to Germany, then Holland, and finally settle in Canada.

The economic contrast of communist Hungary to that of North American capitalism brought about the realization of the vast value that we improperly discard in our country's waste system. During the early 1990s on their many scavenging trips through dumpers, Tom and his dad found working color televisions (something which they never had access to in Budapest), stereos, couches, entire vinyl record collections, and other valuable goods.

It was this fascination with discarded value in combination with Tom's love of entrepreneurship that led him to found

TerraCycle (www.terracycle.com) during his freshman year at Princeton University.

Tom has personally won more than 100 awards for entrepreneurship; he blogs for the *New York Times,* Treehugger, the Huffington Post, and a number of other major websites; and in 2007 he published his first book, *Revolution in a Bottle.* Tom is also the star of the National Geographic Channel TV show *Garbage Moguls.*

About TerraCycle

TerraCycle was founded by Tom Szaky in 2002 with the initial focus to build an organic fertilizer made entirely from waste. TerraCycle's resulting "worm poop" was made by feeding food waste to worms and brewing a tea from their resulting excrement. The product was a highly effective organic fertilizer, packaged in used soda bottles and sold in major retail stores across North America. To collect the bottles for the fertilizer, the team at TerraCycle created an online system (the "Brigades") whereby consumers could send in soda bottles with free shipping for each piece of waste collected, along with a small donation to a school or charity of their choice.

Within years the system became so successful that TerraCycle approached major global manufacturers to collect their waste streams with similar methods. Within 10 years of its founding, TerraCycle expanded its operations to 24 countries around the world, from Australia to Argentina, employing more than 120 associates and engaging over 50 million people in outsmarting waste.

TerraCycle's platform relies on a combination of various collection methods, ranging from industrial collection of latex gloves and other safety equipment in pharmaceutical factories across the United Kingdom and Canada to consumer collection of cookie wrappers from schools and churches across the United Kingdom. Visit www.terracycle .com to learn more.

All of the collected waste is either reused (millions of cell phones are refurbished every year), upcycled (tens of millions of flexible packages are sewn together into pencil cases), or recycled (billions of juice pouches are melted into plastic wood and trash cans), with nothing needing to be burned or buried. These circular solutions emulate nature's timeless processes—where the waste of one organism is the food for another.

TerraCycle has been awarded more than 200 awards for its business model and accomplishments, has been featured in more than 20,000 articles around the world, and works in partnership with most major global manufacturing groups (including Unilever and Kraft) as well as some of the world's biggest waste management companies.

TerraCycle has found that solving the garbage problem, even if it may come at a cost to manufacturers, retailers, government, and even consumers, is something that resonates on a global level with people, corporations, and most importantly our planet.

Berrett–Koehler
Publishers

Berrett-Koehler is an independent publisher dedicated to an ambitious mission: *Creating a World That Works for All.*

We believe that to truly create a better world, action is needed at all levels—individual, organizational, and societal. At the individual level, our publications help people align their lives with their values and with their aspirations for a better world. At the organizational level, our publications promote progressive leadership and management practices, socially responsible approaches to business, and humane and effective organizations. At the societal level, our publications advance social and economic justice, shared prosperity, sustainability, and new solutions to national and global issues.

A major theme of our publications is "Opening Up New Space." Berrett-Koehler titles challenge conventional thinking, introduce new ideas, and foster positive change. Their common quest is changing the underlying beliefs, mindsets, institutions, and structures that keep generating the same cycles of problems, no matter who our leaders are or what improvement programs we adopt.

We strive to practice what we preach—to operate our publishing company in line with the ideas in our books. At the core of our approach is stewardship, which we define as a deep sense of responsibility to administer the company for the benefit of all of our "stakeholder" groups: authors, customers, employees, investors, service providers, and the communities and environment around us.

We are grateful to the thousands of readers, authors, and other friends of the company who consider themselves to be part of the "BK Community." We hope that you, too, will join us in our mission.

A BK Currents Book

This book is part of our BK Currents series. BK Currents books advance social and economic justice by exploring the critical intersections between business and society. Offering a unique combination of thoughtful analysis and progressive alternatives, BK Currents books promote positive change at the national and global levels. To find out more, visit www.bkconnection.com.

Berrett–Koehler
Publishers

A community dedicated to creating
a world that works for all

Dear Reader,

Thank you for picking up this book and joining our worldwide community of Berrett-Koehler readers. We share ideas that bring positive change into people's lives, organizations, and society.

To welcome you, we'd like to offer you a free e-book. You can pick from among twelve of our bestselling books by entering the promotional code **BKP92E** here: http://www.bkconnection.com/welcome.

When you claim your free e-book, we'll also send you a copy of our e-newsletter, the *BK Communiqué*. Although you're free to unsubscribe, there are many benefits to sticking around. In every issue of our newsletter you'll find

- A free e-book
- Tips from famous authors
- Discounts on spotlight titles
- Hilarious insider publishing news
- A chance to win a prize for answering a riddle

Best of all, our readers tell us, "Your newsletter is the only one I actually read." So claim your gift today, and please stay in touch!

Sincerely,

Charlotte Ashlock
Steward of the BK Website

Questions? Comments? Contact me at bkcommunity@bkpub.com.